YOGA: RESEARCH INSIGHTS

Yoga: Research Insights

Edited by:

Shiv Sanjeevi Sripathi

Delve Publishing

www.delvepublishing.com

Yoga: Research Insights

Shiv Sanjeevi Sripathi

Delve Publishing
224 Shoreacres Road
Burlington, ON L7L 2H2
Canada
www.delvepublishing.com
Email: orders@arclereducation.com

ISBN: 978-1-77469-160-1 (Hardcover)

DECLARATION

Some content or chapters in this book are open access copyright free published research work, which is published under Creative Commons License and are indicated with the citation. We are thankful to the publishers and authors of the content and chapters as without them this book wouldn't have been possible.

ABOUT THE EDITOR

ShivSanjeevi Sripathi completed his Masters in Biotechnology from Mumbai University in 2008. He was awarded for academic excellence in both his Bachelors and Masters for securing second rank in Mumbai University in 2006 and first rank in his college: Kishinchand chellaram College. For his Masters he secured first rank in his college KET's V.G.Vaze College. He qualified CSIR and NET and and TOEFL in September 2008. He then worked on a stem cell project at the Specialized Centre for Cell Based Therapy (SCCT),KEM Hospital at Mumbai on a project entitled, : Isolation & detection of stem cells from Human Umbilical cord/ amniotic membrane" following which he worked at Junior Research Fellow at Microbiology & Cell Biology Department, Indian Institute of Science, Bangalore from August 2009 to November 2011. The field of work involved cloning of cell wall genes and transcription factors in *E.coli & M.smegmatis*. As a writer, he has authored and co-authored 25 books with Delve and Arcler International Press. He loves to read and share on interesting aspects of life sciences in books. In his free time he loves to travel and explore ancient customs and traditions.

TABLE OF CONTENTS

LIST OF CONTRIBUTORS

ShivSanjeevi Sripathi

Pooja Swami Sahni
National Resource Centre for Value Education in Engineering, Indian Institute of Technology Delhi, Noida, India

Kamlesh Singh
National Resource Centre for Value Education in Engineering, Indian Institute of Technology Delhi, Noida, India
Department of Humanities and Social Sciences, Indian Institute of Technology Delhi, New Delhi, India,

Nitesh Sharma
National Resource Centre for Value Education in Engineering, Indian Institute of Technology Delhi, Noida, India,

Rahul Garg
National Resource Centre for Value Education in Engineering, Indian Institute of Technology Delhi, Noida, India
Department of Computer Science and Engineering, Indian Institute of Technology Delhi, New Delhi, India
Amar Nath and Shahsi Khosla School of Information Technology, Indian Institute of Technology Delhi, New Delhi, India

Surabhi Gautam
Lab for Molecular Reproduction and Genetics, Department of Anatomy, All India Institute of Medical Sciences, New Delhi, India

Manoj Kumar
Lab for Molecular Reproduction and Genetics, Department of Anatomy, All India Institute of Medical Sciences, New Delhi, India

Uma Kumar
Department of Rheumatology, All India Institute of Medical Sciences, New Delhi, India

Rima Dada
Lab for Molecular Reproduction and Genetics, Department of Anatomy, All India Institute of Medical Sciences, New Delhi, India

Jennifer Huberty
Arizona State University, Tempe, USA.

Mariah Sullivan
Arizona State University, Tempe, USA.

Jeni Green
Arizona State University, Tempe, USA.

Jonathan Kurka
Arizona State University, Tempe, USA.

Jenn Leiferman
Colorado School of Public Health, Denver, USA.

Katherine Gold
University of Michigan Medical School, Ann Arbor, USA.

Joanne Cacciatore
Arizona State University, Tempe, USA.

Ramaprabhu Vempati
Department of Physiology, All India Institute of Medical Sciences, New Delhi, India

Ramesh Lal Bijlani
Department of Physiology, All India Institute of Medical Sciences, New Delhi, India
Sri Aurobindo Ashram, New Delhi, India

Kishore Kumar Deepak
Department of Physiology, All India Institute of Medical Sciences, New Delhi, India

Sana Jarraya
Research Unit, High Institute of Sport and Physical Education, University of Sfax, Sfax, Tunisia

Matthias Wagner
Department of Sport Science, Bundeswehr University Munich, Neubiberg, Germany

Mohamed Jarraya
Research Unit, High Institute of Sport and Physical Education, University of Sfax, Sfax, Tunisia

Florian A. Engel
Department Movement and Training Science, Institute of Sport and Sport Science, Heidelberg University, Heidelberg, Germany

Sergio Elías Hernández
Department of Ingeniería Industrial, Universidad de La Laguna, Tenerife, Spain

José Suero
Centro de Salud Jazmín, Sermas, Madrid, Spain

Alfonso Barros
Department of Psychology, Universitat Jaume I, Castellón, Spain

José Luis González-Mora
Department of Fisiología, Universidad de La Laguna, Tenerife, Spain

Katya Rubia
Institute of Psychiatry, Psychology and Neuroscience, King's College London, London, United Kingdom

Anis Sfendla
Department of Biology, Faculty of Sciences, Abdelmalek Essaâdi University, Tetouan, Morocco

Petter Malmström
Department of Health Sciences, University West, Trollhättan, Sweden

Sara Torstensson
Department of Health Sciences, University West, Trollhättan, Sweden

Nóra Kerekes
Department of Health Sciences, University West, Trollhättan, Sweden

LIST OF ABBREVIATIONS

AC	Attention Control
ACPA	Anti-citrullinated protein antibodies
ACTH	Adrenocorticotropic hormone
ADHD	Attention-Deficit/Hyperactivity Disorder
AIIMS	All India Institute of Medical Sciences
APA	American Psychiatric Association
AQOL	Asthma Quality of Life
ARI	Acute respiratory infection
BDNF	Brain-derived neurotrophic factor
BLI	Bronchial lability index
BMI	Body mass index
BSI	Brief Symptom Inventory
CFA	Confirmatory factor analysis
CG	Control group
CNS	Central nervous system
CRH	Corticotrophin-releasing hormone
CSF	Cerebrum Spinal Fluid
DASS	Depression, anxiety, & stress
DBP	Diastolic blood pressure
DST	Department of Science and Technology
EEG	Electroencephalography
EIB	Exercise-induced bronchoconstriction
ER	Exercise-resistant
ERQ	Emotion Regulation Questionnaire
ES	Exercise-sensitive
FBG	Fasting blood glucose
FS	Fatigue scale
FVC	Forced vital capacity

FWHM	Full-width at half-maximum
GET	Graded exercise therapy
GLM	Generalized Linear Model
GM	Grey matter
GMV	Grey Matter Volume
GSI	General Severity Index
HPA	Hypothalamic-pituitary-adrenal
HRV	Heart rate variability
IHC	Integral Health Clinic
MAAS	Mindful Attention Awareness Scale
MBSR	Mindfulness-based stress reduction
MD	Moderate dose
ME/CFS	Myalgic encephalomyelitis/chronic fatigue syndrome
MID	Minimal Important Difference
MNI	Montreal Neurological Institute
NEPSY	Neuropsychological Assessment
NHANES	National Health and Nutrition Examination Survey
NNT	Number needed to treat
OCIM	Osher Center for Integrative Medicine
PAD	Peptidyl arginine deiminase
PE	Physical education
PEFR	Peak expiratory flow rate
PGS	Perinatal Grief Scale
PHQ-9	Patient Health Questionnaire-9
PSQI	Pittsburgh Sleep Quality Index
PTSD	Post traumatic stress disorder
POMS	Profile of Mood States
QOL	Quality of life
RA	Rheumatoid arthritis
RCT	Randomized controlled trial
RF	Rheumatoid factor
SBP	Systolic blood pressure
SCS	Self-Compassion Scale
SD	Standard deviation

STAI	State-Trait Anxiety Inventory
STC	Stretch and tone control group
SVYASA	Swami Vivekananda Yoga Anusandhana Samsthana
SYM	Sahaja Yoga Meditation
T2DM	Type 2 diabetes mellitus
TC	Total Cholesterol
TICAMs	Tary and alternative medicines
TIV	Total Intracranial Volume
TLR	Toll-like receptors
UCSF	University of California, San Francisco
VBM	Voxel-based morphometry
VLPFC	Ventrolateral prefrontal cortex
WM	White matter
YB	Yogic Breathing
YBLI	Yoga-based lifestyle intervention

PREFACE

All the orthodox systems of Indian Philosophy have one goal in view, the liberation of the soul through perfection. The method is by Yoga.

- Swami Vivekananda.

Derived from the Sanskrit word meaning to join ("yuj"), yoga looks at merging the consciousness of an individual with that of the universal consciousness. The text *Taittiriya Upanishad* elucidates the five layers of our existence viz. *Annamaya kosha* (physical body), *pranamaya kosha* (vital life force), *Manomaya kosha* (mind), *Vijnanamaya kosha* (intellect) and *Anandamaya kosha* (bliss). The sincere practice of yoga takes one to this final layer where one feels bliss leading to mind-body-breath union and health.

Current research is highlighting the benefits of yoga in a host of conditions. This ranges from handling stress, depression, anxiety to overcoming trauma and addressing the management of a slew of diseases such as diabetes, hypertension, irritable bowel syndrome, multiple sclerosis, etc. According to Sengupta, *International journal of preventive medicine*, yoga also contributes to an improvement in the psychological status of cancer patients, lowered physical pain and lesser preterm labor in pregnant women, and an improvement of serum lipid markers in heart patients (health persons too). There have also been reports of improved cognition, emotional handling and alertness in children and the elderly not to mention an improvement in the behavior of prison inmates who underwent a yoga program.

"Yoga does not just change the way we see things, it transforms the person who sees."

— B.K.S Iyengar

Diseases are either *Adhija* (originated by mind) or *Anadhija* (non-stress-related). The former can be *Samanya* (ordinary) that are what we call today as lifestyle noncommunicable diseases, attributed to mental conflicts. *Sara adhija vyadhi* are what we call congenital diseases. Yoga looks at quelling intense emotions such as anger and hate that arise in the *Vijnanamaya kosha* (intellect) to thereby quell ailments and their manifestations.

This book is Part 1 of Yoga: research insights where the benefits offered by this approach are outlined using research articles. It starts off with a chapter on handling COVID-19 (that originated on the last day of 2019 and is still causing waves across the globe). This is followed by how yoga helps manage diabetes-a current issue that affects many and how yoga is useful for other conditions such as chronic fatigue, HIV and inflammation.

The ever-growing benefits offered by yoga are being unearthed by research each passing day. Continued research and well-designed trials can facilitate the incorporation of this ancient approach to help quell diseases and their manifestations across the world.

Note: This book presents insights into yoga to help manage several health conditions. It should not be considered a prescription. Please consult your healthcare provider and if you are interested please learn yoga from a certified organization/instructor.

"Yoga is not a religion. It is a science, science of well-being, science of youthfulness, science of integrating body, mind and soul."

— *Amit Ray, Yoga and Vipassana: An Integrated Life Style*

Chapter 1

Yoga for COVID-19: What does Research Say?

(Shivsanjeevi Sripathi)

CONTENTS

1.1 THE "LOCKDOWN"

In addition to targeting people across the globe, COVID-19 has had multiple impacts. Researchers Ettman and team (2020) reported the first nationally representative study of the prevalence of depressive symptoms in the US. The difference was more than 3-fold higher during the pandemic (March 31, 2020, to April 13, 2020) vs. that of before the COVID-19 pandemic! Of 1441 participants, 27.8% reported depressive symptoms during the time period vs. 8.5% prior to the pandemic. The odds of depression were more in those with less than $5000 in household savings by a factor of 1.5. Table 1 below illustrates the higher incidence of depression during the COVID-19 pandemic vs. that before the pandemic showing its impact on mental health (Ettman et al, 2020: *This is an open-access article distributed under the terms of the CC-BY license, which permits unrestricted use, distribution, and reproduction in any medium. You are not required to obtain permission to reuse this article content, provided that you credit the author and journal*):

Characteristic	Before COVID-19[a]			During COVID-19[b]		
	No. (%)[c]		*P* value[e]	No. (%)[c]		*P* value[e]
	Total (n=5065)	Depression symptoms (n=458)[d]		Total (n=1441)	Depres-sion symp-toms (n=382)[d]	
Sex						
Men	2477 (48.6)	181 (6.9)	.02	723 (48.1)	149 (21.9)	<.001
Women	2588 (51.4)	277 (10.1)		718 (51.9)	233 (33.3)	
Age, y						
18-39	1704 (37.8)	149 (9.0)	.62	619 (38.0)	219 (38.8)	<.001
40-59	1542 (34.2)	144 (8.5)		462 (32.4)	113 (26.8)	
≥60	1819 (28.0)	165 (7.9)		360 (29.7)	50 (14.9)	
Race/ethnicity						

Non-Hispanic White	1790 (62.9)	186 (8.4)	.01	933 (62.9)	225 (26.5)	.32
Non-Hispanic Black	1176 (11.1)	97 (8.4)		143 (11.9)	36 (24.2)	
Hispanic	1155 (15.8)	109 (8.4)		255 (16.6)	84 (34.0)	
Non-Hispanic Asian	674 (5.3)	26 (4.4)		36 (3.1)	8 (23.1)	
Other or multiple	270 (4.8)	40 (16.0)		74 (5.6)	29 (34.4)	
Education						
<High school	958 (11.0)	117 (13.3)	.004	65 (9.8)	22 (29.2)	<.001
High school or GED	1292 (28.4)	129 (9.3)		274 (27.9)	85 (35.0)	
Some college	1636 (30.5)	158 (9.8)		637 (27.8)	186 (32.0)	
≥College	1173 (30.1)	52 (4.7)		465 (34.5)	89 (18.3)	
Marital status						
Married	2404 (53.4)	146 (5.5)	<.001	712 (47.8)	124 (18.3)	<.001
Widowed, divorced, or separated	1093 (18.5)	156 (13.9)		247 (18.4)	75 (31.5)	
Never married	870 (18.7)	91 (10.5)		344 (24.2)	130 (39.8)	
Living with partner	450 (9.4)	42 (10.7)		138 (9.7)	53 (37.7)	
Household income, $						
≤19999	868 (12.9)	130 (16.8)	<.001	246 (19.8)	116 (46.9)	<.001
20000-44999	1319 (24.0)	133 (10.1)		357 (25.8)	109 (31.1)	
45000-74999	887 (19.8)	61 (8.3)		357 (25.1)	83 (23.3)	
≥75000	1354 (43.4)	68 (4.8)		447 (29.3)	67 (16.9)	
Household savings, $						
≤4999	NA	NA	NA	577 (43.2)	227 (40.4)	<.001
≥5000	NA	NA		819 (56.8)	146 (19.3)	
COVID-19 stressor score[f]						

Low	NA	NA	NA	450 (30.7)	64 (15.5)	<.001
Medium	NA	NA		545 (37.5)	132 (25.1)	
High	NA	NA		443 (31.9)	185 (42.9)	
Household size, mean (SD)	3.1 (0.6)	NA	NA	3.2 (0.6)	NA	NA

Abbreviations: COVID-19, coronavirus disease 2019; GED, general education diploma; NA, not available.

[a]Before COVID-19 estimates derived from the National Health and Nutrition Examination Survey (NHANES) from 2017 to 2018. Missing data in the NHANES sample included 637 participants with missing household income, 248 participants with missing marital status, and 6 participants with missing education level.

[b]During COVID-19 estimates derived from the COVID-19 and Life Stressors Impact on Mental Health and Well-being (CLIMB) study collected from March 31 to April 13, 2020. Missing data in the CLIMB study sample included 34 participants with missing household income, 45 participants with missing household savings, and 3 pariticiapnts with missing COVID-19 stressor score.

[c]Frequencies are unweighted. Percentages are weighted. Categories may not add up to total number owing to missing data. Percentages may not add up to 100 owing to rounding.

[d]Defined as Patient Health Questionnaire–9 score of 10 or greater.

[e]Two-tailed $\chi 2$ analysis conducted for significance testing.

[f]Calculated from stressor summation (range, 0-13); stratified as low (score, 0-2), medium (score, 3-4), and high (score, 5-13).

Further, the "***lockdown***" impacted economic factors and livelihoods in addition to sleep issues and inhibited physical activity (Bartos et al, 2021).

1.2 THE BASIS: YOGA FOR ARI

This chapter casts light on instances of how yoga played a role in addressing various aspects of Coronavirus disease 2019 (COVID-19). To start off, the MEPARI (Meditation or Exercise to Prevent Acute Respiratory Infection) -2 Trial explored the role of meditation to address acute respiratory infection (ARI). The trial (Clinicaltrials.gov NCT01654289) entailed the utility of mindfulness-based stress reduction (MBSR) in participants over 8-weeks.

The incidence and the severity of ARI (viruses analyzed by PCR) were lower for the participants who underwent MBSR vs. the controls from 2012 to 2016, with 4 annual cohorts. The global severity of ARI (as observed by the area-under-the-time-severity-curve) was 326 for the controls as opposed to 256 for the MBSR participants. Of 390 participants, the controls reported 105 days of missing work because of ARI vs. 73 days in the MBSR set while viruses were identified in 72 ARI episodes in the control and 63 episodes in the MBSR group. Further, the scores of depression, stress and overall mental health were between in the group that meditated. The authors presented that "*While flu shots confer protection only against influenza, the evidence presented here suggests that mindfulness and exercise training may reduce ARI illness in general, regardless of etiological agent*" (Barrett et al, 2018).

Figure 1 below outlines the outcomes of ARI (acute respiratory infection); ***TOP***: Primary outcome (© 2018 Barrett et al. *This is an open-access article distributed under the terms of the Creative Commons Attribution License, which permits unrestricted use, distribution, and reproduction in any medium, provided the original author and source are credited*):

Outcome	**Exercise (n = 137)**	**Meditation (n = 138)**	**Control (n = 138)**	**Control vs Exercise[1,2]**	**Control vs Meditation[1,2]**
Total # ARIs	120	112	134		
# ARIs per person, mean (95% CI)	0.88 (0.72,1.03)	0.81 (0.66, 0.97)	0.97 (0.80,1.15)	0.10 (-0.14, 0.33) p = 0.42	0.16 (-0.07, 0.39) p = 0.17
Incidence rate (# ARIs per person-year), (95% CI)	1.2 (0.3, 6.0)	1.1 (0.2, 5.8)	1.3 (0.3, 6.0)	0.08 (-0.16, 0.31)	0.15 (-0.08, 0.38)
# People with at least one ARI, n (%)	84 (61%)	74 (54%)	82 (59%)	-0.03 (-0.22, 0.16) p = 0.75	0.10 (-0.10, 0.29) p = 0.33
Total days of ARI Illness	1010	1045	1210		
# days of ARI illness, mean (95% CI)	7.4 (5.7, 9.1)	7.6 (5.6, 9.6)	8.8 (6.8, 10.7)	1.4 (-1.2, 4.0) p = 0.33	1.2 (-1.6, 4.0) p = 0.65
ARI global severity, mean (95% CI)	224 (161, 287)	256 (176, 335)	326 (240, 412)	102 (-4, 208) p = 0.21	70 (-46, 186) p = 0.22

ARI-related missed work days (total)	81	73	105		
ARI-related missed work (hours per person) mean (95% CI)	3.8 (1.9, 5.7)	4.0 (2.4, 5.6)	5.0 (2.9,7.1)	1.2 (-1.6, 4.0) $p = 0.13$	1.0 (-1.6, 3.6) $p = 0.48$
Total # of medical visits	21	22	24		
# Visits to primary care clinic	11	9	19		
# Visits to specialty care clinic	1	4	2		
# Visits to urgent care	4	9	2		
# Visits to hospital/emergency department	4	0	1		
# Visits to other	0	0	0		
Total # of medications used	266	300	381		
Medication use (per ARI), mean (95% CI)					
Mean # prescription medications	0.1 (0.0, 0.1)	0.2 (0.1, 0.2)	0.2 (0.1, 0.3)	0.12 (0.01, 0.23) $p = 0.10$	0.04 (-0.09, 0.17) $p = 0.53$
Mean # OTC medications	2.1 (1.8, 2.5)	2.6 (2.2, 3.0)	2.7 (2.4, 3.0)	0.6 (0.1, 1.0) $p = 0.002$	0.1 (-0.4, 0.6) $p = 0.38$
Mean # all medications	2.2 (1.8, 2.6)	2.8 (2.3, 3.2)	2.9 (2.6, 3.2)	0.7 (0.2, 1.2) $p = 0.001$	0.1 (-0.4, 0.7) $p = 0.37$
ARI economic cost[3] (per person) ($), mean (95% CI)	$119 ($62, $175)	$140 ($83, $197)	$163 ($95, $231)	$44 ($-48, $136) $p = 0.19$	$23 ($-72, $118) $p = 0.18$

Biomarkers[4] (per ARI episode), median (IQR)	91	90	119		
hsCRP (pg/mL)	4.2 (1.5, 9.1)	5.7 (1.4, 12.1)	4.2 (1.2, 12.8)	-0.1 (-1.2, 1.0)[6] p = 0.475	-0.1 (-1.6, 0.6)[6] p = 0.145
Neutrophils (pg/mL)	6.0 (1.0, 26.0)	4.0 (1.0, 25.0)	8.0 (1.0, 38.0)	0.0 (-2.0, 2.0)[5]	1.0 (0.0, 4.0)[5]
IL-6 (serum) (pg/mL)	2.8 (1.9, 4.7)	3.5 (1.8, 5.6)	2.7 (1.5, 4.7)	-0.2 (-0.7, 0.4) [6] p = 0.545	-0.5 (-1.2, 0.1)[6] p = 0.275
IL-6 (nasal) (pg/mL)	25 (5, 91)	29 (6, 81)	16 (4, 67)	-3.7 (-13.7, 0.9)[6] p = 0.135	-3.0 (-14.4, 1.1)[6] p = 0.185
IL-8 (pg/mL)	728 (319, 1450)	472 (231, 886)	551 (296, 1342)	-95 (-253, 52)[6] p = 0.035	104 (-9, 229)[6] p = 0.565
IP-10 (pg/mL)	257 (173, 391)	298 (171, 444)	232 (164, 350)	-25 (-63, 11)[6] p = 0.025	-40 (-89, -1)[6] p = 0.015
Virus identification: # ARIs4, n (%)	91 (76%)	88 (79%)	118 (88%)		
# ARIs found to have any virus present, n (%)	64 (70%)	63 (72%)	72 (61%)	-0.09 (-0.21, 0.04) p = 0.16[1]	-0.11 (-0.23, 0.03) p = 0.12[1]
Specific Virus, n (%)[7]	68	65	76		
Adenovirus (AV)/Bocavirus (BoV)	1 (1.1%)	1 (1.4%)	2 (1.7%)		
Coronavirus (CoV)	12 (13%)	15 (17%)	16 (14%)		
Enterovirus (EV)/Rhinovirus (HRV)	43 (47%)	28 (32%)	41 (35%)		
Influenza	1 (1%)	10 (11.4%)	9 (7.6%)		
MPV/RSV/PIV	11 (12.1%)	11(12.5%)	8 (6.8%)		

IQR = interquartile range, CI = confidence interval, AUC = area under the curve, ARI = acute respiratory infection, hsCRP = high sensitivity C-reactive protein, IL = interleukin, IP = interferon gamma-induced protein, OTC = over the counter, MPV = human metapneumovirus, RSV = respiratory syncytial virus, PIV = human parainfluenza virus.

[1] Between-group comparison p-values comes from: a) Wald Chi-Square test in negative binomial regression for # of ARIs per person; b) Kruskal-Wallis test comparing nonparametric Wilcoxon scores for continuous outcomes (ARI duration, global severity); and c) Wald Chi-square test in logistic regression (for any ARI and any virus).

[2] Between group differences are computed as control group minus treatment group with 95% confidence intervals.

[3] Economic costs were calculated by summing the estimated salary lost from missing work, cost of medical provider visits, and cost of medications used.

[4] Lab visits were within first 72 hours of ARI. Reasons for missing data include: lab closed on weekends, participant unable to get in, etc.

[5] Comparisons are for between group change from baseline (reference group: Control). Neutrophils were not collected at baseline, so no change from baseline could be computed. See Table 1 for baseline biomarkers values.

[6] Hodges-Lehmann estimation for shift in location, with 95% CIs.

[7] More than one virus was identified in 10 separate ARIs. In the EX group, one ARI yielded both BoV and CoV, two ARIs had CoV and HRV, and one ARI had EV/HRV and PIV. In the MBSR group, one ARI yielded EV/HRV and CoV, and one ARI had EV/HRV and PIV (parainfluenza virus). In the control group, one ARI yielded HRV and BoV, one ARI had MPV (metapneumovirus) and CoV, one ARI had RSV (respiratory syncytial virus) and HRV, and one ARI yielded CoV and BoV.

BOTTOM: Secondary outcomes ():

	Exercise				Meditation				Control			
	November/December	January	February/March	April/Exit[1]	November/December	January	February/March	April/Exit[1]	November/December	January	February/March	April/Exit[1]
Self-Reported Instruments (between-group comparison of change from baseline)												
MAAS Mindful Attention	4.4**	4.4**	4.5**	4.5**	4.3**	4.4**	4.4*	4.4**	4.2	4.2	4.3	4.3
	(2.6, 5.8)	(2.6, 5.8)	(2.7, 6.0)	(2.6, 6.0)	(2.9, 5.6)	(2.8, 5.6)	(2.8, 5.9)	(2.8, 5.9)	(2.7, 5.7)	(2.6, 5.8)	(2.7, 6.0)	(2.7, 5.9)
SF12: Physical Health	51.7	51.7	51.6	51.9	51.2	51.6	51.4	50.5	52.0	51.6	50.8	51.5
	(34.8, 69.2)	(33.5, 69.8)	(31.4, 70.2)	(32.2, 70.9)	(34.8, 69.3)	(35.7, 67.5)	(35.3, 66.4)	(31.4, 71.7)	(36.0, 68.1)	(34.1, 69.2)	(33.2, 68.4)	(33.9, 69.2)
SF12: Mental Health	47.6	47.7	47.6	47.8	48.0	49.2*	49.3*	49.7**	45.5	45.5	46.2	45.9
	(23.5, 67.5)	(24.5, 66.6)	(24.8, 67.7)	(26.2, 65.6)	(27.7, 63.3)	(29.0, 62.1)	(28.0, 64.5)	(25.2, 66.5)	(24.5, 66.4)	(24.8, 66.2)	(25.6, 66.9)	(25.1, 66.6)
SPS Social Support	82.4	82.1	83.8	83.1	82.5	82.8	83.6	83.8	81.4	81.6	82.2	82.6
	(60.8, 101.9)	(61.7, 101.4)	(61.2, 103.2)	(62.2, 103.1)	(60.1, 102.6)	(60.4, 102.7)	(60.8, 103.7)	(61.5, 103.7)	(59.3, 103.5)	(61.6, 101.5)	(60.9, 103.5)	(62.6, 102.7)
PSS10 Perceived Stress	13.0**	12.3**	12.4**	13.3	13.0*	12.4*	12.3**	11.6**	14.0	13.8	13.6	13.4
	(0.6, 27.5)	(0.2, 27.3)	(0.5, 26.8)	(-0.5, 27.4)	(2.5, 25.5)	(2.6, 24.9)	(1.3, 25.9)	(0.8, 26.0)	(1.6, 26.5)	(1.0, 26.5)	(1.1, 26.2)	(0.6, 26.2)
PSQI Sleep Quality	5.4**	5.5*	5.4*	5.3**	5.5	4.9	5.2	5.1	6.0	5.8	5.6	5.8
	(-0.2, 12.2)	(-0.5, 12.0)	(-0.9, 12.2)	(-0.4, 12.0)	(-0.3, 12.4)	(-0.9, 12.4)	(-1.2, 12.5)	(-0.1,11.8)	(-0.6, 12.6)	(-0.7, 12.2)	(-0.7, 12.0)	(-1.3, 13.0)
GPAQ Exercise[2]	1560**	1380**	1210**	1440**	700	480	680	980	670	540	600	1080
	(860, 2770)	(600, 2320)	(600, 2210)	(640, 2940)	(120, 1692)	(0, 1260)	(180, 1560)	(240, 2460)	(160, 1560)	(40, 1560)	(4, 1740)	(360, 2740)

MSES Mindful Self-Efficacy	99.0**		99.0*	99.9**	100.7**		101.4**	102.6**	94.8		94.8	95.1
	(63.3, 126.4)		(63.8, 125.8)	(66.6, 123.6)	(65.3, 124.3)		(65.4, 124.1)	(63.8, 126.5)	(64.5, 125.1)		(62.0, 127.5)	(63.0, 127.3)
ESES Exercise Self-Efficacy	104.6**		104.3**	104.4**	92.7		96.3*	94.3*	86.0		84.3	84.6
	(13.8, 158.2)		(11.2, 157.4)	(13.4, 155.8)	(14.8, 157.3)		(12.6, 156.0)	(8.9, 160.2)	(17.8, 154.3)		(11.2, 157.4)	(13.3, 155.9)
PANAS Positive	34.8		34.5	34.9	35.1		34.8	35.8	33.6		32.7	33.2
	(17.9, 49.3)		(17.5, 47.9)	(17.6, 48.7)	(18.8, 48.4)		(20.3, 45.1)	(20.1, 46.2)	(18.6, 48.5)		(18.1, 47.3)	(18.2, 48.1)
PANAS Negative	17.9		17.7	17.9	17.7		17.5	17.4	18.7		19.0	17.9
	(5.8, 31.6)		(6.4, 31.6)	(5.6, 30.3)	(8.2, 29.3)		(7.9, 30.1)	(7.1, 28.8)	(7.5, 30.0)		(6.5, 31.4)	(7.2, 28.7)
PHQ9 Depressive Symptoms	4.3*		4.1*	4.1	4.4		3.9	3.6	5.3		4.9	4.5
	(-2.2, 12.8)		(-2.8, 12.5)	(-2.2, 11.2)	(-2.2, 12.8)		(-2.7, 12.4)	(-2.9, 12.0)	(-3.2, 13.7)		(-3.2, 12.9)	(-3.2, 12.2)
Stanford Presenteeism	24.1		24.3	24.2	23.8		23.8	24.0	23.1		23.0	23.6
	(13.2, 32.9)		(13.9, 32.2)	(14.0, 33.3)	(13.5, 32.6)		(14.6, 31.4)	(14.4, 32.9)	(13.7, 32.4)		(13.8, 32.2)	(15.2, 32.1)
Feeling Loved[2]	366.5		363.0	370.0	377.0		372.5	380.0	365.0		362.0	361.0
	(338.5, 389.5)		(339.0, 387.0)	(342.0, 388.0)	(342.0, 390.0)		(345.0, 390.0)	(350.0, 390.0)	(330.0, 381.0)		(340.0, 385.0)	(340.0, 385.0)
Body Measurements (between-group comparison of change from baseline)												
BMI (kg/m^2)	29.3*		29.3	29.2	29.7		29.7		29.2		28.9	29.1
	(15.1, 43.3)		(14.6, 43.3)	(14.6, 43.6)	(14.0, 44.4)		(14.0, 44.4)		(16.1, 42.3)		(16.3, 41.6)	(15.9, 42.3)

Systolic BP[3] **(mmHg)**	124 (99, 157) /122 (95, 157)		120	121	122		122		126		124	125
			(97, 152)	(95, 154)	(93, 159)		(93, 159)		(98, 154)		(93, 156)	(93, 157)
Diastolic BP[3] **(mmHg)**	77 (61, 96) / 75 (58, 95)		76	76	75		75		77		77	77
			(61, 93)	(62, 93)	(58, 95)		(58, 95)		(60, 93)		(61, 92)	(61, 93)
Biomarkers[2] (within-group comparison of change from baseline)												
IL-6 (se-rum) (pg/mL)[2]	1.7		1.6		1.6		1.7		1.5		1.	
	(1.0, 3.0)		(0.9, 2.5)		(0.9, 3.2)		(0.9, 2.6)		(1.0, 2.6)		(1.1, 3.0)	
IL-6 (nasal) (pg/mL)[2]	1.8		1.6$^{\#}$		1.7$^{\#\#}$		1.6$^{\#}$		1.3$^{\#\#}$		1.6$^{\#\#}$	
	(0.8, 2.9)		(0.8, 2.8)		(0.8, 3.9)		(0.9, 2.6)		(0.6, 2.8)		(0.8, 3.6)	
IL-8 (pg/mL)[2]	214$^{\#\#}$		227$^{\#}$		238$^{\#\#}$		258$^{\#\#}$		255$^{\#\#}$		248$^{\#\#}$	
	(127, 354)		(158, 414)		(144, 356)		(157, 424)		(140, 446)		(155, 390)	
IP-10 (pg/mL)[2]	139$^{\#\#}$		145$^{\#}$		140		149		148		147	
	(116, 172)		(111, 190)		(116, 183)		(114, 192)		(121, 205)		(123, 192)	
hsCRP (pg/mL)[2]	1.7		1.3		1.7		1.3$^{\#\#}$		1.6		1.4	
	(0.7, 3.3)		(0.6, 3.2)		(0.8, 4.8)		(0.7, 3.3)		(0.7, 4.4)		(0.6, 3.9)	
HbA1c (pg/mL)[2]	5.5		5.5		5.6$^{\#\#}$		5.5		5.6$^{\#\#}$		5.5	
	(5.3, 5.8)		(5.3, 5.9)		(5.3, 5.8)		(5.2, 5.8)		(5.4, 5.9)		(5.3, 5.8)	

Mean (95% confidence interval) unless otherwise noted.

See Table 1 for baseline values.

Abbreviations: SPS = social provisions scale, PANAS = positive and negative affect schedule, PSS = perceived stress scale, PSQI = Pittsburg sleep quality index, MAAS = mindfulness attention awareness scale, MSES = mindfulness self-efficacy scale, ESES = exercise self-efficacy scale, GPAQ = global physical activity questionnaire, BP = blood pressure, HbA1c = hemoglobin A1c, hsCRP = high sensitivity C-reactive protein, IL = interleukin, IP = interferon gamma-induced protein.

[1]Exit visits occurred late May or early June.

[2]Median (interquartile range).

[3] Blood pressures were collected in November at the flu shot visit and at the December in-clinic follow-up visit. All other measures were collected at only one month out of the two months listed.

Between group change from baseline comparison for that visit (reference group: Control group change from baseline):

* < 0.05,

** < 0.01 (nonparametric Wilcoxon scores in the Kruskal-Wallis test).

Biomarkers Only: Within treatment group change from baseline comparison for that visit (reference group: baseline for that treatment group):

< 0.05,

< 0.01 (nonparametric Wilcoxon Signed Rank Test.

Looking at COVID-19, a 2020-published research article explored the bibliometric trends of traditional, integrative, complementary and alternative medicines (TICAMs) globally to address COVID-19. Figure 2 below is illustrative of the publications assessed that reported the use of one article using yoga (Ng, 2020: *Open-access: This article is licensed under a Creative Commons Attribution 4.0 International License, which permits use, sharing, adaptation, distribution and reproduction in any medium or format, as long as you give appropriate credit to the original author(s) and the source, provide a link to the Creative Commons licence, and indicate if changes were made. The images or other third party material in this article are included in the article's Creative Commons licence, unless indicated otherwise in a credit line to the material. If material is not included in the article's Creative Commons licence and your intended use is not permitted by statutory regulation or exceeds the permitted use, you will need to obtain permission directly from the copyright holder. To view a copy of this licence, visit http://creativecommons.org/licenses/by/4.0/. The Creative Commons Public Domain Dedication waiver (http://creativecommons.org/publicdomain/zero/1.0/) applies to the data made available in this article, unless otherwise stated in a credit line to the data*):

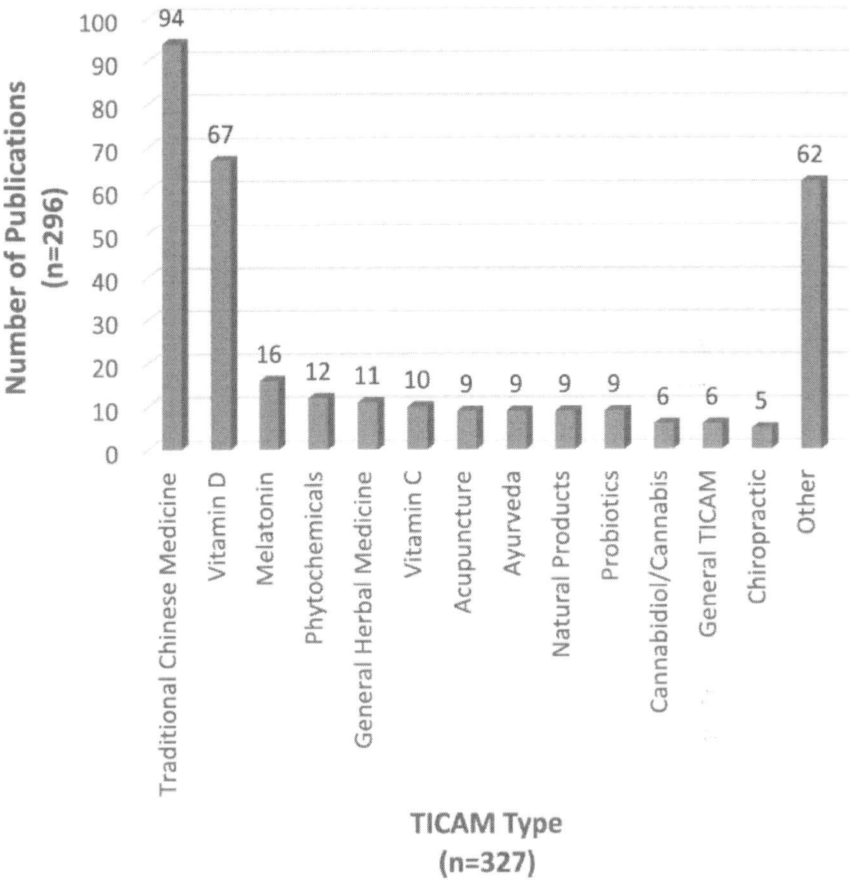

1.3 CASE STUDY: BENEFITS OF YOGA IN MUSIC STUDENTS DURING THE LOCKDOWN

The effect of the CRAFT [*Consciencia* (i.e., consciousness), *Relajación* (i.e., relaxation), *Atención* (i.e., attention), *Felicidad* (i.e., happiness), and *Transcendencia* (i.e., transcendence)] program was assessed in a higher conservatory of music during the lockdown and online teaching in terms of addressing distress. The participants were 18-39 years (40 from the CRAFT program and 53 controls) of the Royal Conservatory of Music Victoria,

Eugenia. The 23-h program entailed the use of hatha yoga, postures and meditation for meeting *Consciousness*; pranayama and asanas for *Relaxation*; pranayama, mantras with raja yoga (*dharana*, *dhyana*, and *samadhi*) for *Attention* to lead to *Happiness* and *Transcendence*.

The health and wellbeing details are depicted below in Figure 3 (© 2021 Bartos, Funes, Ouellet, Posadas and Krägeloh. *This is an open-access article distributed under the terms of the* ***Creative Commons Attribution License (CC BY)****. The use, distribution or reproduction in other forums is permitted, provided the original author(s) and the copyright owner(s) are credited and that the original publication in this journal is cited, in accordance with accepted academic practice. No use, distribution or reproduction is permitted which does not comply with these terms*):

Perceived outcomes	Total sample			Filtered sample[a]		
	CRAFT program group ($n = 40$)	Control group ($n = 53$)	p	CRAFT program group ($n = 31$)	Control group ($n = 31$)	p
Severe COVID-19 setbacks	4 (10%)	5 (9%)	0.93	5 (16%)	2 (7%)	0.23
Health and well-being Changes	29 (72%)	32 (61%)	0.22	22 (71%)	23 (74%)	0.78
Positive changes	6 (15%)[b]	3 (6%)[b]	0.13	4 (13%)[b]	3 (10%)[b]	0.69
Mental–emotional issues	21 (52%)[b]	27 (51%)[b]	0.88	17 (55%)[b]	20 (64%)[b]	0.43
Anxiety	13 (32%)[b]	10 (19%)[b]	0.13	10 (32%)[b]	9 (29%)[b]	0.78
Stress	13 (32%)[b]	14 (26%)[b]	0.52	10 (32%)[b]	12 (39%)[b]	0.60
Fear, worry, and despondency	6 (15%)[b]	6 (11%)[b]	0.60	6 (19%)[b]	5 (16%)[b]	0.74
Physical issues	14 (35%)[b]	10 (19%)[b]	0.08	11 (35%)[b]	6 (19%)[b]	0.15
Tiredness	6 (15%)[b]	3 (6%)[b]	0.13	6 (19%)[b]	2 (6%)[b]	0.13
Sleep disturbances	4 (10%)[b]	3 (6%)[b]	0.43	3 (10%)[b]	2 (6%)[b]	0.64
10-point VAS health states (1–10)						
Physical state	6.80 ± 1.83	6.92 ± 1.97	0.76	6.97 ± 1.83	6.90 ± 2.02	0.90
Mental and emotional state	6.23 ± 1.69	5.98 ± 2.2	0.55	6.03 ± 1.68	5.87 ± 2.41	0.76
Sleep quality	5.93 ± 2.4	5.91 ± 2.50	0.97	6.06 ± 2.46	6.00 ± 2.57	0.92
Engagement in health practices	37 (92%)	31 (58%)	**< 0.001**	29 (93%)	17 (55%)	**< 0.001**
Yoga/meditation practices	35 (95%)[b]	6 (19%)[b]	**< 0.001**	27 (93%)[b]	2 (12%)[b]	**< 0.001**

CRAFT	25 (68%)[b]	0 (0%)[b]	NA	21 (72%)[b]	0 (0%)[b]	NA
Other yoga/ meditation	13 (35%)[b]	6 (19%)[b]	0.15	7 (24%)[b]	2 (12%)[b]	0.31
Other practices	18 (49%)[b]	29 (93%)[b]	< **0.001**	15 (52%)[b]	15 (88%)[b]	**0.012**
Alexander technique	2 (5%)[b]	8 (26%)[b]	**0.018**	2 (7%)[b]	4 (23%)[b]	0.11
Exercise	17 (46%)[b]	23 (74%)[b]	**0.018**	14 (48%)[b]	12 (71%)[b]	0.141
Unspecified	0 (0%)[b]	1 (3%)[b]	0.27	0 (0%)[b]	0 (0%)[b]	NA
Perceived benefits from practices	29 (78%)	16 (52%)	**0.020**	23 (79%)	7 (41%)	**< 0.01**
Yoga/meditation	26 (70%)[b]	1 (3%)[b]	< **0.001**	20 (69%)[b]	0 (0%)[b]	< **0.001**
CRAFT	19 (51%)[b]	0 (0%)[b]	< **0.001**	16 (55%)[b]	0 (0%)[b]	< **0.001**
Other yoga/ meditation	10 (27%)[b]	1 (3%)[b]	**0.008**	5 (17%)[b]	0 (0%)[b]	0.07
Other practices	12 (32%)[b]	15 (48%)[b]	0.18	10 (34%)[b]	6 (35%)[b]	0.96
Alexander technique	0 (0%)[b]	5 (16%)[b]	**0.011**	0 (0%)[b]	3 (18%)[b]	**0.02**
Exercise	12 (32%)[b]	9 (29%)[b]	0.76	10 (34%)[b]	3 (18%)[b]	0.22
Unspecified	0 (0%)[b]	1 (3%)[b]	0.27	0 (0%)[b]	0 (0%)[b]	NA

Values are mean ± SD with respective p-values computed with one-way ANOVA, or n (%) with respective p-values computed with χ^2. Bold font denotes significance. All displayed percentages refer to the total number of participants within group except for the variables related to the practices implemented and their perceived benefits whereby they refer to the total number of proactive participants within group; [a]Filtered sample after excluding participants that reported previous exposure with yoga/ Meditation other than in the CRAFT program before the lockdown and in the CRAFT program during the academic years 2017-2019. [b]Percentages from non-mutually exclusive categories. NA = Not applicable.

Few benefits as reported by the participants are illustrated in Figure 4 below (© 2021 Bartos, Funes, Ouellet, Posadas and Krägeloh. *This is an open-access article distributed under the terms of the* ***Creative Commons Attribution License (CC BY)****. The use, distribution or reproduction in other forums is permitted, provided the original author(s) and the copyright owner(s) are credited and that the original publication in this journal is cited, in accordance with accepted academic practice. No use, distribution or reproduction is permitted which does not comply with these terms*):

Practices implemented	Exemplar responses from proactive participants					
	Craft program group ($N_{TS} = 37$; $N_{FS} = 29$)	Beneficial effect		Control group ($N_{TS} = 31$; $N_{FS} = 17$)	Beneficial effect	
		N_{TS} (%)	N_{FS} (%)		N_{TS} (%)	N_{FS} (%)
Yoga/Meditation	"I have practiced Meditation. I have learnt it in the Emotional intelligence class and it has been an important element to bear the situation. The changes have been relevant. Management of anxiety, stress, interfamily relationships, time management. It has been an essential help in multiple aspects."	16 (43)	12 (41)	"Yoga learnt through video lessons on YouTube. Honestly, I have noticed certain improvement from the first practice though I have been doing it for not a long time to feel more important beneficial effects."	1 (3.2)	1 (6)
	"The various techniques learnt in the Mindfulness subject such as the physical Protocol, Meditations, breath control helped me live with the current situation and see it with a neutral perspective."					
Yoga/Meditation + Other	"Sports made me liberate accumulated tensions and feel much better with myself. I meditated a lot and thanks to the subjects of Mindfulness and Emotional intelligence I am able to manage and self-regulate my emotions, I am more conscious of my thoughts, and my sate in general."	11 (30)	9 (31)	"Guided meditations that helped me relax during anxious moments. Sports helped me feel my body stronger."	3 (9.7)	0 (0)
				"Meditation (audio from authorized courses), cardio, push-ups, abs, squats. All of this has helped me bear the effects I have described but I also believe that in a lockdown state like this there is no one who can be healthy regardless of what they do."		

	"Every day I did physical activity that helps me to relax/calm down physically, mentally, and emotionally. I also practiced meditation but not as much as I would have liked. It was difficult as it requires a prior state of concentration and calmness. However, when I applied it I have noticed outstanding changes."					
Other	"Online training with a personal training through Instagram. It has been very positive because it was helping me to unleash my energy and to put my body in operation".	2 (5)	2 (7)	"Only doing sports has helped me keep active, feel more energic and be more positive about looking at things."	12 (38.7)	6 (40)
				"I used the 'conscious rest' practice from the ergonomics subject when I was overwhelmed and could not take it anymore. Resting for 5–10 min and let myself go without thinking of anything gave me tranquility and relaxation, which I needed it."		
	"Sports five times a week and I am very proud. I have not experienced a lot of change but it makes me feel well".					

N_{TS} = Number of proactive participants in the total sample; N_{FS} = Number of proactive participants in the filtered sample after excluding participants that reported previous exposure with yoga/Meditation other than in the CRAFT program before the lockdown and/or in the CRAFT program during the academic years 2017–2019.

The participants of the CRAFT program reported an augmented engagement in activities to boost their wellbeing: 95%, employed yoga and/or meditation practices vs. 19% of their counterpart controls! Given the vital involvement of yoga in this program, the authors attribute the approach of yoga to tackle the root of an ailment and self-caring to account for the participating students to account for their health (Bartos et al, 2021).

1.4 CASE STUDY: SUCCESSFULLY TREATING A HIGH-RISK CASE

The first report of a combination of yoga and Ayurveda to treat COVID-19 (high-risk case) was published in 2021 in the Journal of Medical Case Reports by researchers Mishra and team. The recipient was a 55-year-old man with hypertension and type 2 diabetes for which he was on relevant allopathic drugs. He had been administered nitazoxanide 500 mg, doxycycline 100 mg, pantoprazole 40 mg, vitamin C, multivitamin (methylcobalamin, pyridoxine and folic acid) with levocetirizine 5 mg upon diagnosis of COVID-19 for ten days.

Based on the test reports, the current authors recommended the following yoga practices delivered online and through a video:

- *Sukshma Vyayama* (joint movements)
- Breathing exercises
- *Asanas* (*Parshva Sukhasana*, *Sukhasana* twist, *Utthana Mandukasana*, *Ardha Ushtrasana*, *Meru Vakrasana*, *Ardha Halasana*—with one leg folded, *Anantasana*, *Pawan Muktasana*), *Shavasana*,
- *Pranayama* (breathing: *Anulom-Vilom*, *Bhramari*, *Udgeet*)
- *Dhyana* (mindful breathing)
- *Yoga Nidra* (guided relaxation) was also recommended to quell the anxiety observed.

This yoga was coupled with the following Ayurvedic medicines as depicted below in Table 2 (Mishra et al, 2021: *Open-access* This article is licensed under a Creative Commons Attribution 4.0 International License, which permits use, sharing, adaptation, distribution and reproduction in any medium or format, as long as you give appropriate credit to the original author(s) and the source, provide a link to the Creative Commons licence, and indicate if changes were made. The images or other third-*party material in this article are included in the article's Creative Commons licence, unless indicated otherwise in a credit line to the material. If material is not included in the article's Creative Commons licence and your intended use is not permitted by statutory regulation or exceeds the permitted use, you will need to obtain permission directly from the copyright holder. To view a copy of this licence, visit http://creativecommons.org/licenses/by/4.0/. The Creative Commons Public Domain Dedication waiver (http://creativecommons.org/publicdomain/zero/1.0/) applies to the data made available in this article, unless otherwise stated in a credit line to the data*):

Common name	Botanical name
Giloy Ghanvati	
Giloy	*Tinospora cordifolia*
Ashwagandha vati	
Ashwagandha	*Withania somnifera*
Pathyadi Kwath (pravahi)	
Harad	*Terminalia chebula*
Baheda	*Terminalia bellerica*
Amla	*Emblica officinalis*
Neem chhal	*Azadirachta indica*
Giloy	*Tinospora cordifolia*
Haldi	*Curcuma longa*
Diabecon	
Shilajeet	*Asphaltum punjabia-num*
Meshashringi/Gudmara/Madhunashini	*Gymnema sylvestre*
Peetashala/Vijaysar/Indian Kino Tree	*Pterocarpus marsu-pium*
Turmeric	
Haldi	*Curcuma longa*

Spiritual practices (mantra chanting) and dietary modifications were also incorporated. The overall plan and response of the patient are documented below in table 3: *FBS* fasting blood sugar, *PPBS* postprandial blood sugar, *RBS* random blood sugar, *RT-PCR* real-time reverse transcription polymerase chain reaction (Mishra et al, 2021: *Open-access* This article is licensed under a Creative Commons Attribution 4.0 International License, which permits use, sharing, adaptation, distribution and reproduction in any medium or format, as long as you give appropriate credit to the original author(s) and the source, provide a link to the Creative Commons licence, and indicate if changes were made. The images or other third party material in this article are included in the article›s Creative Commons licence, unless indicated otherwise in a credit line to the material. If material is not included

Day	Date	FBS (mg/dl)	PPBS (mg/dl)	RBS (mg/dl)	Follow-up details
0	1/7/2020	–	–	–	Generally the blood sugar values were FBS>200 mg/dl, PPBS>250 mg/dl while the patient was taking allopathic medication for the same; PPBS was reported to be 230 mg/dl on 30/6/2020
1	2/7/2020	192	–	–	Symptoms reported/observed: difficulty in speech, weakness, sometimes feeling of bad smell and taste, heaviness in abdomen (due to gas formation), uneasiness in body, less appetite
2	3/7/2020	169	–	–	Improvement observed; approximately 25% relief reported in all symptoms; lightness in the body reported
3	4/7/2020	190	206	–	Doing Yoga (twice daily); feeling good, with improvement in all symptoms
4	5/7/2020	160	206	–	Approximately 50% relief reported in all symptoms; lightness in the body reported
5	6/7/2020	155	175	–	No weakness; normal appetite; taste and smell almost normal; approximately 75% relief reported. Second RT-PCR done
6	7/7/2020	140	–	110	No sore throat; no weakness; normal taste and smell; almost no other problem except mild heaviness in abdomen; approximately 80% relief reported
7	8/7/2020	140	–	110	No problem except mild heaviness in abdomen
8	9/7/2020	130	230	–	Second RT-PCR report received on 8/07/2020 was positive; slight change made in the prescription: (i) Diabecon 1 tablet in the morning and 2 tablets in the evening (previously 1 tablet in the evening); (ii) advised to take (ajwain + saunf + dhaniya) powder (1/2 spoon) with lukewarm water after meals
9	10/7/2020	110	150	–	Improvement in all problems (including heaviness in abdomen)

10	11/7/2020	110	150	–	No problem except mild heaviness in abdomen; dose of Diabecon changed to 2 tablets twice daily (patient advised to start the revised dose from 12/7/2020)
11	12/7/2020	130	140	–	No problem except mild heaviness in abdomen; started Diabecon 2 tablets twice daily
12	13/7/2020	99	112	–	No problem except mild heaviness in abdomen. Third RT-PCR done
13	14/7/2020	108	112	–	No problem except mild heaviness in abdomen
14	15/7/2020	120	165	–	No problem except mild heaviness in abdomen
15	16/7/2020	108	140	–	RT-PCR test inconclusive as per the report; no problem except mild heaviness in abdomen
16	17/7/2020	90	140	–	No problem except mild heaviness in abdomen
17	18/7/2020				No problem except mild heaviness in abdomen
18	19/7/2020				No problem except mild heaviness in abdomen
19	20/7/2020				Sample given for RT-PCR test; the patient's allopathic doctor reduced the dosage of his diabetes medication, that is glimepiride changed from 2 mg to 1 mg; no problem except mild heaviness in abdomen
20	21/7/2020				Diabecon changed to Diabecon DS 1 tablet twice daily; no problem except mild heaviness in abdomen
21	22/7/2020				RT-PCR negative; no problem except mild heaviness in abdomen

The patient feedback is as follows:

"*The patient expressed immense satisfaction and a sense of confidence. He was amazed and delighted to find that besides aiding in a speedy recovery from the symptoms of COVID-19, including weakness, anxiety and depression, Yoga and Ayurveda had managed to control his blood sugar level as well. The normal sugar levels observed by the end of the treatment were astounding given that regular allopathic medication could not achieve this. The patient was so convinced about the efficacy of Yoga and Ayurveda that he suggested his spouse (who was also tested positive for COVID-19 and had been pursuing only allopathic medication) seek Ayurvedic consultation. Most importantly, he is determined to adopt Yoga in his lifestyle*" (Mishra et al, 2021: *Open-access* This article is licensed under a Creative Commons Attribution 4.0 International License, which permits use, sharing, adaptation,

 Thus, this study of successfully addressing the various symptoms of a patient while also boosting the immunity and mental health opens the avenue of employing yoga as a vital part of addressing COVID-19, the virus that stopped the world!

1.5 FINAL POINTS

According to Nagendra (2020), a video of breathing exercises was sent to patients at Milano, Italy hospitalized for COVID-19 infection-associated acute respiratory distress. While the period of March 17, 2020, and March 20, 2020 saw 1000 people visiting this pilot video, a renowned cardiac surgeon in Italy using an oxygen mask in an intensive care unit utilized the technique and said "we have reached scientific evidence that this simplified protocol sent by you is effective and we intend to disseminate to the overall Scientific Community" (Nagendra, 2020).Bushell et al (2020) reviewed the suitability of yoga and meditation as adjuvant approaches to handle COVID-19. The levels of melatonin are augmented in those who practice yoga /meditation-this becomes significant given that this molecule is known to target viruses while also exerting antioxidant and immune-modulating attributes. The ability of yoga to augment the levels of β-defensin and HBD-2: antimicrobial peptides in the respiratory epithelium become noteworthy for COVID-19 given the targeting of the tissue by the virus.

Overall, "*The authors propose that certain complementary practices may be helpful adjunctive* means of treating and/or preventing SARS-CoV-2 infection[1,9] and helping to reduce severity of COVID-19 disease, including its collateral effects and sequelae" (Bushell et al, 2020).

The stage is set to employ yoga to help quell the damage by COVID-19!

1.6 REFERENCES

1. Ettman CK, Abdalla SM, Cohen GH, Sampson L, Vivier PM, Galea S. Prevalence of Depression Symptoms in US Adults Before and During the COVID-19 Pandemic. *JAMA Netw Open.* 2020;3(9):e2019686. doi:10.1001/jamanetworkopen.2020.19686
2. Bartos LJ, Funes MJ, Ouellet M, Posadas MP and Krägeloh C (2021) Developing Resilience During the COVID-19 Pandemic: Yoga and Mindfulness for the Well-Being of Student Musicians in Spain. *Front. Psychol.* 12:642992. doi: 10.3389/fpsyg.2021.642992
3. Barrett B, Hayney MS, Muller D, Rakel D, Brown R, Zgierska AE, et al. (2018) Meditation or exercise for preventing acute respiratory infection (MEPARI-2): A randomized controlled trial. PLoS ONE 13(6): e0197778. https://doi.org/10.1371/journal.pone.0197778
4. Ng, J.Y. Global research trends at the intersection of coronavirus disease 2019 (COVID-19) and traditional, integrative, and complementary and alternative medicine: a bibliometric analysis. *BMC Complement Med Ther* 20, 353 (2020). https://doi.org/10.1186/s12906-020-03151-8
5. Mishra, A., Bentur, S.A., Thakral, S. *et al.* The use of integrative therapy based on Yoga and Ayurveda in the treatment of a high-risk case of COVID-19/SARS-CoV-2 with multiple comorbidities: a case report. *J Med Case Reports* 15, 95 (2021). https://doi.org/10.1186/s13256-020-02624-1
6. Nagendra H R. Yoga for COVID-19. Int J Yoga [serial online] 2020 [cited 2021 Apr 28];13:87-8. Available from: https://www.ijoy.org.in/text.asp?2020/13/2/87/283658
7. William Bushell, Ryan Castle, Michelle A. Williams, Kimberly C. Brouwer, Rudolph E. Tanzi, Deepak Chopra, and Paul J. Mills. Meditation and Yoga Practices as Potential Adjunctive Treatment of SARS-CoV-2 Infection and COVID-19: A Brief Overview of Key Subjects. The Journal of Alternative and Complementary Medicine.Jul 2020.547-556.http://doi.org/10.1089/acm.2020.0177

Chapter

2

Yoga to Handle Diabetes

(Shivsanjeevi Sripathi)

CONTENTS

2.1 THE SCENARIO

According to the global and regional diabetes prevalence estimates for 2019 and projections for 2030 and 2045, diabetes accounted for four million mortalities in the world and global health expenditure of 727 billion US$ in 2017! Further, globally a little less than 50% of the global population has diabetes with these figures expected to escalate by 25% by 2030 and 51% by 2045 (Saeedi et al, 2019).

According to the Canada Clinical Practice Guidelines Expert Committee (2018), Type 1 diabetes is when the pancreas is incapacitated to synthesize insulin and type 2 diabetes is when the produced insulin doesn't meet the demand or the body cannot use the synthesized hormone. Diagnosing diabetes during pregnancy gives it the name gestational diabetes while there are other forms due to genetics or drug usage.

Figure 1 below is illustrative of the status of diabetes (Martínez-Castelao et al: © 2015 by the authors; licensee MDPI, Basel, Switzerland. *This article is an open-access article distributed under the terms and conditions of the Creative Commons Attribution license (http://creativecommons.org/licenses/by/4.0/)*:

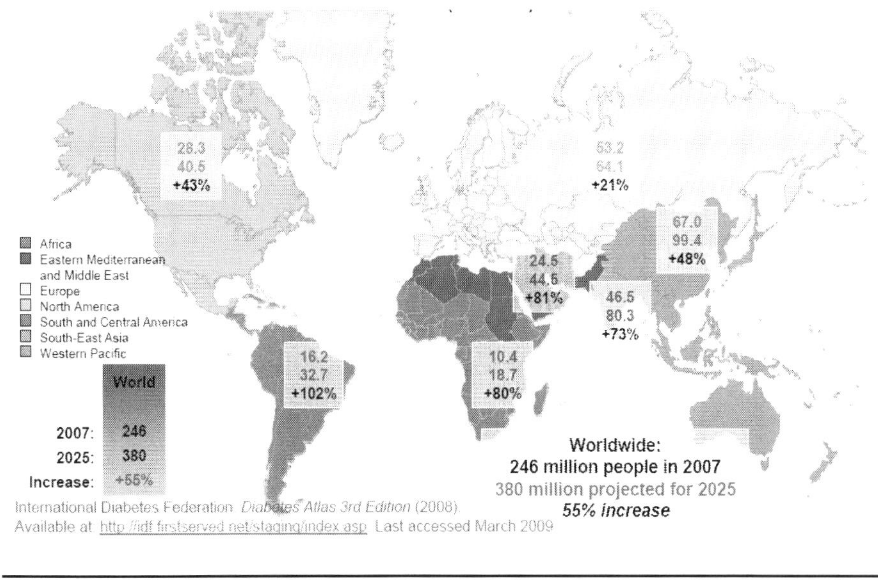

While diet, medication and exercise are vital to managing diabetes, this chapter presents instances of research showing the effect of yoga in managing this condition.

2.2 ADDRESSING THE RISK OF DIABETES

A pilot study entailed the random assignment of 23 adults to either an educational group or a yoga-practicing group for three months. The recruitment criteria were: Age: 45 and 65 years; Family history of diabetes; Minimum exercise or non-exercisers; demonstrative of the risk factors-affected fasting glucose (<125 mg/dl); prehypertension (systolic BP/ diastolic BP: 120–139/80–89 mmHg); overweight/obese [body mass index (BMI) >25 but <45 kg/m^2]; or aberrant cholesterol (total cholesterol >200 mg/dl).

The educational control group received general health-education content addressing the various risk factors along with data on adherence to diet and exercise every two weeks. The yoga group involved training on a Vinyasa-style yoga program inclusive of postures (*Asanas*) and regulated breathing (*Pranayamas*). Table 1 below documents the details of the participants with headache and depression as major complaints; 87% of them displayed aberrant cholesterol profiles and one-third had hypertension (Yang et al, 2011: *Copyright © 2011 Kyeongra Yang et al. This is an open-access article distributed under the Creative Commons Attribution License, which permits unrestricted use, distribution, and reproduction in any medium, provided the original work is properly cited*):

	n (%)	**Mean +- SD**
Age (years)		51.7 4.9
Gender		
Male	2 (8.7)	
Female	21 (91.3)	
Race		
White	19 (82.6)	
Black	2 (8.7)	
Hispanic	1 (4.3)	
Native American	1 (4.3)	
Level of education (years)		18.0 4.1
Sedentary lifestyle	23 (100)	
Family history of diabetes	23 (100)	
Impaired fasting glucose	2 (8.7)	

Pre-hypertension	7 (30.4)	
Overweight/obese (BMI 30 kg/m^2)	20 (87)	
Abnormal cholesterol levels	20 (87)	
Number of co-morbidities		5.9 3.6

Table 2 below documents the various aspects of the yoga scheme: group sessions were weekly twice with participants expected to practice at home (Yang et al, 2011: *Copyright © 2011 Kyeongra Yang et al. This is an open-access article distributed under the Creative Commons Attribution License, which permits unrestricted use, distribution, and reproduction in any medium, provided the original work is properly cited*):

Warm-up (5–7 min)
Standing, seated or supine; includes all movements of the spine
Sun-salutations (10–12 min)
Full or modified, depending on the participant's needs
Standing poses (12–15 min; poses are held for 3-4 breaths)
Two-legged balances (warrior series, triangle series)
One-legged balances
Seated/kneeling poses and counter-poses (12–15 min; poses are held for 2-3 breaths)
Spinal extension
Hip-opener
Spinal rotation; Spinal flexion
Relaxation (10 min) using deep relaxation pose

Interestingly, the association between systolic blood pressure and attendance in the program was significant. Improvements were documented for weight, systolic BP, diastolic BP, insulin, total cholesterol, triglycerides and exercise self-efficacy as demonstrated below in table 3 (Yang et al, 2011: *Copyright © 2011 Kyeongra Yang et al. This is an open-access article distributed under the Creative Commons Attribution License, which permits unrestricted use, distribution, and reproduction in any medium, provided the original work is properly cited*):

	Yoga (n=12)			**Education (n=10)**			*z*	***P*-value**	**Effect size**
	Baseline	3 months	Change	Baseline	3 months	Change			

	Mean (SD)	Mean (SD)		Mean (SD)	Mean (SD)				
Weight (Ib)	175.5 (30.0)	174.7 (29.8)	−0.79	188.4 (28.8)	189.4 (29.4)	1.02	−1.385	.166	−0.53
BP (mmHg)									
Systolic	119.3 (14.3)	114.1 (7.7)	−5.17	115.7 (15.9)	119.3 (13.9)	3.60	−1.519	.129	−0.62
Diastolic	68.7 (8.3)	69.3 (9.0)	0.58	68.9 (9.5)	72.6 (7.5)	3.70	−0.925	.355	−0.35
Fasting glucose level	90.1 (11.5)	92.8 (11.7)	2.75	91.2 (13.9)	95.8 (10.8)	4.60	−0.396	.692	−0.20
Insulin	10.5 (5.8)	8.1 (5.7)	−2.42	9.8 (5.3)	9.9 (7.4)	0.01	−0.799	.424	−0.34
Lipid panel									
Total cholesterol	209.8 (33.6)	193.3 (32.1)	−16.50	198.8 (46.5)	187.6 (34.1)	−11.20	−0.726	.468	−0.22
Triglycerides	120.5 (41.3)	109.0 (46.6)	−11.50	98.3 (48.1)	114.3 (47.9)	16.00	−1.914	.056	−0.91
LDL-C	130.3 (35.2)	117.8 (33.9)	−12.41	126.1 (46.8)	116.0 (32.2)	−10.10	−0.132	.895	−0.10
HDL-C	60.3 (10.1)	56.6 (12.8)	−3.75	57.0 (15.3)	51.4 (11.8)	−5.60	−0.298	.766	0.24
Exercise self-efficacy	43.9 (23.2)	57.9 (23.0)	13.97	52.5 (30.5)	49.2 (23.7)	−3.33	−1.583	.114	0.80

To quote the authors, "*the results of this pilot study suggest that a yoga program could potentially be a risk reduction option for adults at high risk for type 2 diabetes. Anecdotal comments by study participants revealed that they perceived improvement in their strength, flexibility and balance after practicing yoga* (Yang et al, 2011).

The effect of yoga on individuals with metabolic syndrome or prediabetes was scrutinized in a systematic review and meta-analysis (*reported as the first*) by Ramamoorthi and team (2019). Studies involving yoga to tackle type 2 diabetes mellitus (T2DM) with relevant controls along with relevant health factors between Jan 2002 and Dec 2018 were explored across databases (PubMed, Scopus, Cochrane Library, EBSCO and IndMED). Following the identification of 46,500 studies, filtering resulted in 14 studies of 834 participants as presented in Figure 2 below (© 2019 Ramamoorthi et al: *This is an open-access article distributed under the terms of the Creative Commons Attribution License, which permits unrestricted use, distribution, and reproduction in any medium, provided the original author and source are credited*):

S.No	Author and Year	Country	Participation time	Yoga Asanas	Intervention	No of patients	Age and Sex	Control group	No of patients	Age and Sex	Follow-up period	Outcome measures	Study design	Mean and SD values
1	Corey SM et al. 2014 [81]	USA	NA	Restorative yoga intervention	Yoga participants	88	21–65 yrs	Stretching participants	83	21–65 yrs	One year	Salivary cortisol and Psychosocial measurements	Two-arm RCT	NA
2	Hegde et al. 2013 [49]	India	2007–2008	18 Yoga asanas	Pre-diabetics patients	14	30 and 75 yrs: Male-6: Female-8	Computer generator randomised list	15	30 and 75 yrs Male-8: Female-7	3 months	Oxidative stress, glycemic status, Blood pressure, Anthropometry	RCT	Provided
3	Kanaya Am et al 2014 [72]	USA	2009–2012	Restorative yoga intervention	Patients	88	21–65 yrs: F-65: M-23	Stretching participants	83	21–65 yrs F-59: M-24	48 weeks	fasting and 2-hour glucose, HbA1c, triglycerides, HDL Cholesterol	RCT	Provided
4	Keerthi GS et al. 2017 [66]	India	September 2013 to April 2016.	Asanas and pranayama, Meditation and Relaxation	Pre-diabetics and Diabetics	124	18–45 yrs Prediabetics-M-69/F-55: Diabetics-M-64/F-60	Healthy controls	62	M-34/F-28	12 weeks	Biochemical measures, Anthropometric measures, Bp, plasma glucose	RCT	Provided
5	Lau Caren et al. 2015 [83]	China	May 2010 -Jan 2011	57 Yoga Poses	Yoga Practitioners	87	18 and Above M-34: F-53	Non-Yoga practitioners	88	18 and Above M-30: F-56	12 weeks	Metabolic risk factors, Mets z score, Glucose, BP, triglyceride, Smoking, Body weight	Non-RCT	Provided
6	McDermott KA et al. 2014 [69]	India	Oct-Nov 2004	8 types of Yoga asanas and chanting	Yoga Practitioners	21	Age -NA Male-9 Female-12	Non-yoga practitioners, only physical exercises	20	Age -NA; Male-7 Female-13	8 weeks	Diabetic Risk Factors and Physiological Risk Factors	RCT	Provided
7	Netam et al. 2015 [67]	India	Dec 2011-Dec 2012	Asanas and pranayama	overweight/obese individuals	34	M-21: F-13	Pre-baseline	34	M-21: F-13	one month	IL-6, 25-OH-vitamin D and diabetes risk factors	Non-RCT	Provided

8	Siu et al. 2015 [84]	Hong-kong	Nov 2010 and Aug 2013	34 Yoga asanas	Metabolic syndrome	84	30–80 M-23: F-75	Computer generator ran-domised list	98	30–80 M-24: F-60	One year	Wait, BP, glucose, triglycerides, HDL-C, Heart rate	RCT	Pro-vided
9	Sohl et al 2016 [71]	USA	June 2013 to January 2014	12 Yoga-asanas	Metabolic syndrome patients	33	M-17: F-16	education only	33	M-16; F-17	12 weeks	Biometric and Patient-reported constructs	RCT	Pro-vided
10	Supriya et al. 2017 [73]	China	NA	18 Yoga-asanas	Metabolic syn-drome patients associated with Diabetes & Cardiovascular diseases	52	M-17: F-35	Computer generator ran-domised list	45	M-17: F-28	One year	Blood glucose, triglyc-erides, HDL-C, Waist circumference	RCT	Pro-vided
11	Tyagi A et al. 2014 [68]	India	Nov and Jan 2012	Yoga teachers, Yoga therapists and Yoga active person	Yoga Practi-tioners	NA	18 yrs to 55 yrs	Non-yoga practitioners	NA	18 yrs to 55 yrs	6 months	Anthropometric mea-sures, BP measurement, metabolic measures	RCT	NA
12	Wolff M et al. 2013 [85]	Sweden	1 May 2008 and 31 January 2010	NA	three groups: through Yoga practitioner, Yoga at home, a control group who take mediation from a general practitioner	56	20–80 yrs	Computer generator ran-domised list	27	20–80 yrs	12 weeks	Bp, glucose, HBA1c, Cholesterol	three-arm RCT	Pro-vided
13	Yadhav et al. 2018 [70]	India	Septem-ber 2013 to April 2016.	Asanas and pranaya-ma, Meditation and Relax-ation	Metabolic syndrome patients	130	20–45 years Gender number details not provided	dietary inter-vention	130	20–45 years Gender number details NA	12 weeks	Weight, Height, Body mass, BP, Glucose, Insulin, HDL-C, Insulin, Triglyceride	RCT	Pro-vided
14	Yang K et al. 2011 [86]	USA	NA	Asanas and pranayama	Patients	23	45 and 65 years, M-2; F-21	Pre-baseline	23	45 and 65 years, M-2; F-21	3 months	Weight, Bp, Glucose, Insulin, Lipid panel	Ran-domised Control Trial	Pro-vided

Analyses revealed that yoga augmented the clinical management of several factors associated with prediabates viz. "*fasting blood glucose (FBG) [Standard Mean Difference (SMD -0.064 mg/dL (95% CI -0.201 to 0.074)]; low density lipoprotein (LDL) [SMD-0.090 mg/dL (95% CI -0.270 to 0.090)]; triglycerides [SMD -0.148 mg/dL (95% CI -0.285 to -0.012)]; total cholesterol [SMD -0.058 mg/dL (95% CI -0.220 to 0.104)] and systolic blood pressure [SMD -0.058 mm Hg (95% CI -0.168 to 0.053)]*" (Ramamoorthi et al, 2019).

Figure 3 below documents the impact of yoga on various parameters of prediabetes: A) fasting blood glucose, B) Postprandial blood glucose, C) Glycosylated hemoglobin (HbA1c), D) LDL, E) Triglycerides, F) Total Cholesterol (TC), G) Waist circumferences, H) Body weight, I) Systolic blood pressure (SBP) and J) Diastolic blood pressure (DBP) (© 2019 Ramamoorthi et al: *This is an open-access article distributed under the terms of the Creative Commons Attribution License, which permits unrestricted use, distribution, and reproduction in any medium, provided the original author and source are credited*):

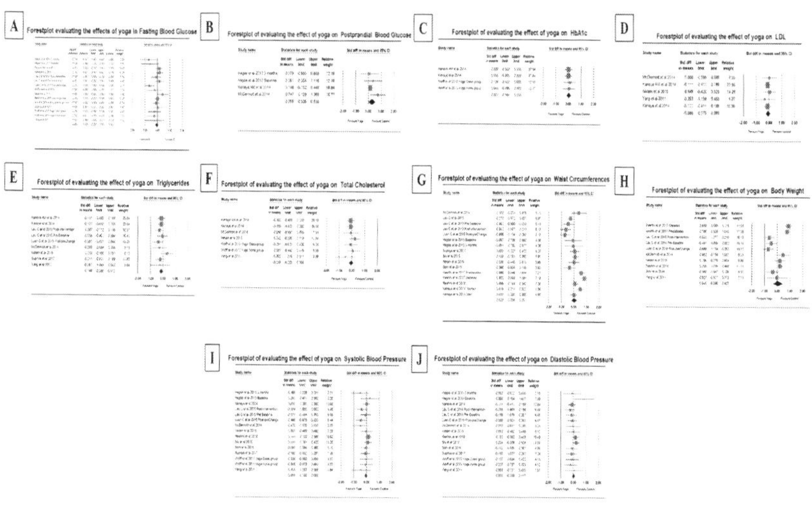

Overall, given the improvement in the parameters, the results are suggestive of looking at yoga to quell T2DM by addressing prediabetic manifestations. Relevant trials can further add weight to open the avenue of utilizing the ancient approach to handle the disease.

2.3 MANAGING DIABETES

The benefits bestowed by physical exercise are manifold, especially in the scenario of diabetes as shown below in figure 4 below (© 2015 Thangasami SR, et al: *This is an open-access article distributed under the terms of the Creative Commons Attribution License, which permits unrestricted use, distribution, and reproduction in any medium, provided the original author and source are credited*):

Structural changes in skeletal muscle
↑ fiber size
↑ percentage of type IIa fibers (Possibly type I fibers)
↑ the capillary density and blood flow
Biochemical changes in skeletal muscle
↑ insulin signaling kinetics (1-phosphatidyl inositol 3-kinase and GLUT4 activity)
↑ non-insulin signaling kinetics (5'-AMP-activated protein kinase activity)
↑ enzymes related to glucose metabolism [hexokinase, glycogen synthase, and key oxidative enzymes (e.g., citrate synthase, aconitase, succinate dehydrogenase)]
↑ myoglobin
Systemic influences of physical activity
↑ oxygen uptake and functional capacity at submaximal and maximal workloads
↑lipoprotein lipase and other key enzymes to improve lipid control
↓ excessive hepatic secretion of glucose and VLDL
Improvements in counter-regulatory hormone levels/activity (e.g. Cortisol, IGF-I)
Improvements in comorbid conditions (e.g., hypertension, visceral obesity, systemic inflammation, dyslipidemia)

Table 1: Potential mechanisms by which exercise enhance glucose homeostasis [21]

While yogic breathing (*pranayama*) allows the mind and body to be linked while managing health ailments, the postures (*asanas*) facilitate balancing and strengthening of the muscles, joints and posture while allowing homeostasis of several systems. The following figure 5 depicts a scheme of pranayama and yoga practices to manage T2DM on the basis of literature by Thangasami, Chandani and Thangasami (2015) (© 2015 Thangasami SR, et al: *This is an open-access article distributed under the terms of the Creative Commons Attribution License, which permits unrestricted use, distribution, and reproduction in any medium, provided the original author and source are credited*):

S.NO	NAME OF ASANA	DURATION	
1	Bhastrika Pranayama	3 - 5 mins per day	
2	Suryanamskar	3 - 7 turns	
3	Kapal- Bhati	5 - 10mins per day	
4	Tadasana (Mountain pose)	¼ minute to one minute	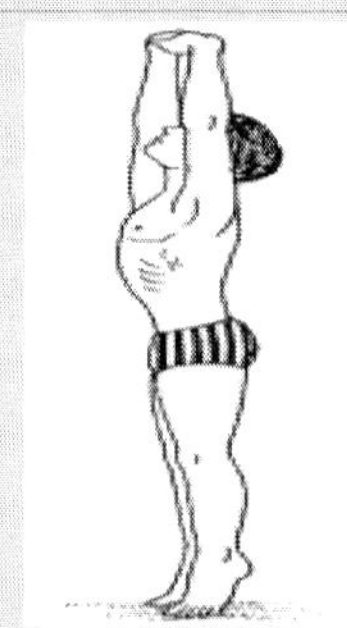
5	Trikonasana (Triangle pose)	¼ minute to one minute	
6	Anulom Viloma (breathing exercise)	5 - 10 mins per day	
7	Bhramari(breathing exercise)	5 times a day	
8	Paschimottanasan (seated forward bend)	¼ minute to one minute	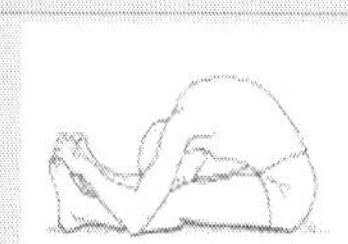 Paschimottanasana (Seated Forward Bend)
9	Bhujangasana (cobra pose)	3 - 7 turns	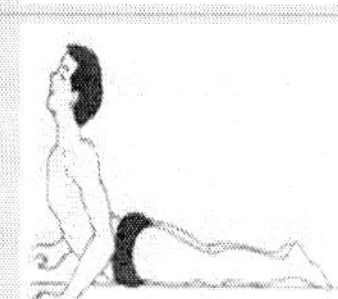 Bhujangasana
10	Shavasana (corpse pose)	3 - 7 turn of each	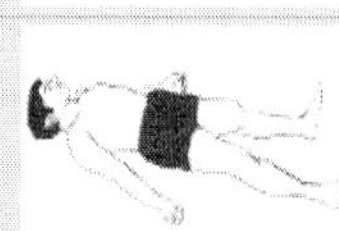
11	Vajrasana (thunderbolt pose)	¼ minute to one minute	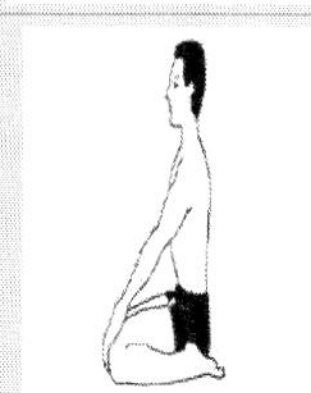

12	Pawanmuktasana (wind relieving pose)	¼ minute to one minute	
13	Padmasana(lotus position)	5 - 15 minutes per day	
14	Sarvangasana(supported shoulder stand pose)	Few seconds	
15	Matsyasana(fish pose)	Few seconds	
16	Sukhasana(crossed leg pleasant pose)	¼ minute to one minute	

Author's note: Please do not consider this as a prescription/remedy. Kindly consult your medical health provider and learn yoga from a qualified instructor if you wish to practice.

2.4 CLINICAL TRIALS

A pilot study (ClinicalTrials.gov Identified NCT00090506) explored the effects of an eight-week yoga intervention vs. control (walking) at the Swami Vivekananda Yoga Anusandhana Samsthana (SVYASA) in Bangalore, India and collaborated with the Osher Center for Integrative Medicine (OCIM) at the University of California, San Francisco (UCSF) based on advertisements. The screening criteria were FBG (fasting blood glucose) ≥5.6 mmol/L, a first-degree relative with T2DM; 30–65 years old and readiness to be part of a yoga program.

Table 4 below depicts the scheme of yoga followed to target stress, augment positivity and manage metabolism (McDermott et al, 2014: *This article is published under license to BioMed Central Ltd. This is an Open-access article distributed under the terms of the Creative Commons Attribution License (http://creativecommons.org/licenses/by/2.0), which permits unrestricted use, distribution, and reproduction in any medium, provided the original work is properly credited. The Creative Commons Public Domain Dedication waiver (http://creativecommons.org/publicdomain/zero/1.0/) applies to the data made available in this article, unless otherwise stated*):

Didactics—10 minutes	
Diabetes causes, complications and lifestyle factors	
Principles, philosophy, and practices related to a yoga based lifestyle program	
Stress response	
Maladaptive behavior and behavior change	
Emotion and coping	
2.	***Pranayama (Breathing exercises)—6 minutes***
	Hands stretch breathing
	Ankle stretch breathing
	Tiger breathing
	Rabbit breathing
3.	***Loosening exercises (any 3)—10 minutes***
	Jogging
	Forward, backward, side bending
	Twisting
	Pavanamuktasana kriya (Supine knees to chest)
	Surya namaskara (3 sun salutations)

4.	***Standing asana (any 3)—8 minutes***
	Padahasthasana (Foot hand)
	Ardhachakrasana (Half moon)
	Trikonasana (Triangle)
	Parshvakonasana (Side angle)
	Ardhakati chakrasana (Half wheel)
	Vrikshasana (Tree)
6.	***Supine asana—8 minutes***
	Sarvangasana (Shoulder stand)
	Halasana (Plough)
	Matsyasana (Fish)
	Pavanamuktasana (Supine knee chest position)
	Naukasana (Boat)
7.	***Prone asana—8 minutes***
	Bhujangasana I, II , III (Cobra)
	Shalabhasana – alternate legs, both (Locust)
	Dhanurasana (Bow)
	Navasana (Boat)
8.	***Sitting asana—8 minutes***
	Paschimottanasana (Seated forward bend)
	Vakrasana/ardhamatsyendrasana (Half twist)
	Ustrasana (Camel)
	Sashankasana (Rabbit)
9.	***Relaxation Shavasana (Corpse) with guided scan—6 minutes***
10	***Chanting 'OM' monosyllables, primordial sounds in Indian philosophy***

The BMI, weight and waist size reduction were conspicuously lowered in those who were part of the yoga group (final number=20) vs. the walking control (final number=18). Over the duration of the study, blood pressure and total cholesterol were evidently lowered in both groups along with an overall improvement in feeling positive in the yoga group. The researchers overall reported the plausibility of employing yoga to address T2DM risk factors inclusive of BMI, weight and waist reduction (McDermott et al, 2014).

Table 5 below documents the effects of the interventions (McDermott et al, 2014: *This article is published under license to BioMed Central Ltd. This is an Open-access article distributed under the terms of the Creative Commons Attribution License (http://creativecommons.org/licenses/by/2.0), which permits unrestricted use, distribution, and reproduction in any medium, provided the original work is properly credited. The Creative Commons Public Domain Dedication waiver (http://creativecommons.org/publicdomain/zero/1.0/) applies to the data made available in this article, unless otherwise stated*):

	Yoga	**Control**	**Diff. (Yoga-Ctrl)**	**P diff.**	**Cohen's d**
	(n = 20)	**(n = 18)**	**[95% CI]**		
Diabetes risk factors					
Fasting BG (mmol/L)	0	−0.3	0.3 [−0.5 to 1.1]	0.43	0.27
Postprandial BG (mmol/L)[a]	−0.6	0.1	−0.7 [−2.4 to 0.9]	0.38	0.37
BMI (kg/m2)	−0.2	0.6	−0.8 [−1.6 to −0.0]	0.05	0.68
Weight (kg)	−0.8	1.4	−2.3 [−4.1 to −0.4]	0.02	0.80
Waist (cm)	−4.2*	0.7	−4.9 [−7.9 to −1.9]	<0.01	1.11
Systolic (mmHg)	−5.3*	−5.4*	0.1 [−4.8 to 5.1]	0.95	0.02
Diastolic (mmHg)	−5.3*	−3.6*	−1.7 [−7.6 to 4.1]	0.55	0.20
Insulin (pmol/L)	−16.7	4.2	−20.9 [−52.2 to 10.4]	0.19	0.46
Insulin resistance	−0.3	0.1	−0.4 [−1.0 to 0.2]	0.17	0.48
LDL (mmol/L)	−0.1	0.0	−0.2 [−0.6 to 0.3]	0.47	0.24
Total cholesterol (mmol/L)	−0.9*	−0.7*	−0.2 [−0.9 to 0.6]	0.67	0.15
Triglycerides (mmol/L)	−0.1	−0.1	0 [−0.6 to 0.6]	0.95	0.02
Psychological factors [b]					
Anxiety	−2.1*	−3.4*	1.3 [−0.8 to 3.4]	0.22	0.42

Depression	−2.5*	−3.2*	0.7 [−1.7 to 3.2]	0.55	0.21
Positive affect	4.8*	4.6	0.2 [−4.9 to 5.2]	0.95	0.02
Negative affect	−5.0*	−8.4*	3.4 [−2.3 to 9.0]	0.24	0.41
Perceived stress	−5.7*	−7.9*	2.2 [−1.3 to 5.8]	0.21	0.43

Another study published in 2017 described as an "open label parallel three armed randomized control trial" (CTRI/2011/12/002227 dated 14/12/2011) looked at glycemic outcomes and yoga in 124 women diagnosed with T2DM in rural Kerala, India. Random assignment of the individuals was done into three groups:

- Control: standard care
- Peer group: Monitoring of the individuals by a peer mentor (who received training on advising on exercise, diet, medicines and social/emotional support)
- The yoga group was inclusive of a twice-weekly-60 min session on "25 min-Surya namaskara-12 steps, 5–7 min-Deep relaxation-Muscle relaxation technique, 15 min Asanas: or yoga postures: Pavanamuktasana (supine), Bhujangasana and Shalabhasana (prone), Ardhamatsyaendrasana (sitting) and 15 min of pranayama. This was to be followed at home too.

The yoga group demonstrated a dip in systolic (6.1 mm: p=.08) and diastolic blood pressure (3.1 mm: p=.03). The dip in FPG (fasting plasma glucose) was 6.5 mg/dl vs. 5.9 mg/dl in peer and 1.7 mg/dl in control while for HbA1c (glycosylated hemoglobin) only the yoga group showed a decrease by 0.5% while the values were increased by 0.5% and 0.19% in the peer and control groups (Sreedevi et al, 2017).

Table 6 below illustrates the changes due to the interventions discussed above (Copyright © The Author(s). 2017: Sreedevi et al, 2017, *Open-access: This article is distributed under the terms of the Creative Commons Attribution 4.0 International License (http://creativecommons.org/licenses/by/4.0/), which permits unrestricted use, distribution, and reproduction in any medium, provided you give appropriate credit to the original author(s) and the source, provide a link to the Creative Commons license, and indicate if changes were made. The Creative Commons Public Domain Dedication waiver (http://creativecommons.org/publicdomain/zero/1.0/) applies to the data made available in this article, unless otherwise stated*): [a] *FPG*: Fasting plasma glucose; [b]HbA1c: Glycosylated Hemoglobin; [c] *BP*: Blood pressure

	Yoga				Peer				Control			
Variable	**No**	**Mean**	**SD**	*p*	**no**	**Mean**	**SD**	*p*	**no**	**Mean**	**SD**	*P*
[a]FPG(mg/dl) Pre	32	166	43.03	0.51	25	169.5	62.78	0.6	30	183.58	72.16	0.9
Post	32	159.5	44.08		25	163.54	69.72		30	181.84	71.77	
[b]Hb A1C% Pre (mmol/mol)	32	9.6 (82)	1.62	0.39	25	9.3 (78)	1.6	0.15	30	9.4 (79)	1.71	0.5
Post	32	9.4 (79)	1.54		25	9.8 (84)	2.26		30	9.6 (81)	1.94	
Adherence Pre	30	5.55	1.88	0.1	21	5.42	1.67	0.58	29	5.03	1.79	0.02
Post	30	6.07	1.49		21	5.61	1.96		29	5.75	1.49	
Systolic[c] BP (mmHg) Pre	32	133.81	20.91	0.02	26	127.07	19	0.55	30	126.06	16.17	0.34
Post	32	127.75	18.62		26	129.46	19.21		30	129.26	17.91	
DiastolicBP (mmHg) Pre	32	84.25	11.74	0.07	26	81.84	9.39	0.66	30	80.2	7.34	0.06
Post	32	81.12	9.18		26	81	10.6		30	83.3	8.5	
Total Cholesterol (mg/dl) Pre	32	218.87	48.81	0.34	25	233.19	40.93	0.32	30	211.37	36.42	0.09
Post	32	225.13	44.76		25	228.24	35.82		30	227.2	55.59	
Waist Hip Ratio Pre	32	0.96	0.045	0.34	27	0.97	0.04	0.02	30	0.98	0.051	0.002
Post	32	0.95	0.053		27	0.94	0.04		30	0.94	0.054	
Body Mass Index (kg/m^2) Pre	32	24.74	3.97	0.09	31	25.54	4.52	0.05	35	24.86	4.07	0.01
Post	32	22.82	7.16		31	21.72	9.36		35	20.32	10.06	

While 90% of the women found that the peer support was useful; 80% of the yoga group participants were regular thus opening up the avenue to employ yoga to manage the distress associated with diabetes.

2.5 FINAL POINTS

Several other research studies conducted on the benefits presented by yoga to manage diabetes are presented below in figure 6 (© 2015 Thangasami

SR, et al: *This is an open-access article distributed under the terms of the Creative Commons Attribution License, which permits unrestricted use, distribution, and reproduction in any medium, provided the original author and source are credited*):

S.no	Author	Design	Population (N)	Intervention	Duration	Outcomes Measured	Results
1	Malhotra V_ [60]	Pre- post design	20 type 2 DM patients	Yoga were performed every day along with medicines in the study group while the control group only with medicines and light excercises.	40 days	The basal blood glucose, nerve conduction velocity of the median nerve was measured and repeated after 40 days of Yogic regime	Control group nerve function parameters deteriorated over the period of study, indicating that diabetes is a slowly progressive disease involving the nerves. Yoga asanas have a beneficial effect on glycaemic control and improve nerve function in mild to moderate Type 2 diabetes with sub-clinical neuropathy.
2	Manjunatha et.al. [48]	Pre-post design	20 healthy young volunteers	Each volunteer performed four sets of asanas in random order for 5 consecutive days each with a 2-day gap between consecutive sets of asanas. The four sets of asanas were : Dhanurasana& matsyendrasana, halasana & vajrasana naukasana & bhujangasana, setubandhasana& pavanamuktasana	4-week	Blood samples were collected on days 4 and 5 of each set of asanas for measurement of glucose and insulin levels before the asanas, within 10 min after performing the asanas, and 30 min after ingestion of 75 g glucose, which in turn was ingested immediately after the second blood sample. A standard 75 g oral glucose tolerance test (OGTT) was also done before and after the study. On the days of the pre-study or post-study OGTT, no asanas were done	Performance of asanas led to increased sensitivity of the B cells of pancreas to the glucose signal. The increased sensitivity seems to be a sustained change resulting from a progressive long-term effect of asanas.
3	Malhotra V_ [52]	Pre- post design	20 NIDDM patients	13 specific Yogaasanas were taught in the study group and the study parameters assessed before and after study period.	40 days	Serum insulin, plasma fasting and one hour postprandial blood glucose levels and anthropometric parameters were measured before and after yoga asanas	At the end of 40 days of performing the asanas, the study participants had a significant decrease in fasting glucose levels, waist-hip ratio and beneficial changes in insulin levels than the previous levels
4	Bijlani RL et al. [50]	Pre- post design	heterogeneous group of 98 patients with hypertension, coronary artery disease, diabetes mellitus, and a variety of other illnesses	yogasanas pranayama (breathing exercises), relaxation techniques	9 days	fasting plasma glucose and serum lipid profile	Fasting plasma glucose, serum total cholesterol, low-density lipoprotein (LDL), VLDL , the ratio of total cholesterol to high density lipoprotein (HDL) cholesterol, and total triglycerides were significantly lower, and HDL cholesterol significantly higher, on the last day of the course compared to the first day of the course.
5	Kristal AR_ [54]	Retrospective analysis	15,500 adults recruited for VITAL cohort study	Physical activity (including yoga) during the past 10 years, diet, height, and weight at recruitment and at ages 30 and 45	10 years	Measures were based on self-reporting, and past weight was retrospectively ascertained.	Regular yoga practice was associated with attenuated weight gain, most strongly among individuals who were overweight. Although causal inference from this observational study is not possible, results are consistent with the hypothesis that regular yoga practice can benefit individuals who wish to maintain or lose weight
6	Chaya MS_ [62]	Comparison study	30 healthy young males	15 healthy young male patients practicing yoga for long termwere compared with 15 young males not practicing yoga		insulin sensitivity was measured in the fasting state by the hyperinsulinaemic-euglycaemic clamp	Long term yoga practice (for 1 year or more) is associated with increased insulin sensitivity and attenuates the negative relationship between body weight or waist circumference and insulin sensitivity.
7	Telles S [55]	Pre- post design	single group of 47 healthy persons	persons were assessed on the first and last day of a yoga and diet change program	6 days	Body mass index (BMI), waist and hip circumferences, mid-arm circumference, body composition, hand grip strength, postural stability, serum lipid profile and fasting serum leptin levels	A 6-day yoga and diet change program decreased the BMI and the fat-free mass. Total cholesterol also decreased due to reduced HDL levels and a decrease in serum leptin levels

8	Kyizom T. [61]	Comparison study	Sixty patients of type 2 diabetes	control group on only conventional medical therapy and yoga-group on conventional medical therapy along with pranayama and yoga-asana	45 days	Basal recordings of P300 and blood glucose were taken at the time of recruitment and second recordings repeated after forty five days for both the groups. P300 was recorded	yoga has a beneficial effect on P300 and thus can be incorporated along with the conventional medical therapy for improving cognitive brain functions in diabetes.
9	Balaji PA [59]	Comparison study	44 type 2 DM patients	22 patients were yoga group and the control group did not practice yoga	3 months	FBS, PPBS, Hb A1c, lipid profile and anthropometric measurements like weight, BMI, Waist-hip ratio were estimated before the starting and at the end of the study period	Yoga group had significant decrease in FBS, Postprandial blood sugar (PPBS), glycosylated hemoglobin (HbA1c), triglycerides and LDL with $P < 0.001$, compared with control group (n = 22).
10	Hegde, Shreelaxmi V. et al. [78]	Controlled clinical trial	123 type 2 diabetic patients aged between 40 and 75 years	Patients were grouped as 60 for yoga and 63 for control	3 months	Malondialdehyde glutathione superoxide dismutase , vitamin C and vitamin E were measured to assess the oxidative stress and antioxidant status. BMI, waist circumference waist-tohip ratio, blood pressure, fasting plasma glucose (FPG), postprandial plasma glucose (PPPG), and HbA1c were analyzed	In comparison with standard care alone, yoga resulted in significant reduction in BMI, glycemic control, and malondialdehyde and increase in glutathione and vitamin C. There were no differences in waist circumference, waist-to-hip ratio, blood pressure, vitamin E, or superoxide dismutase in the yoga group at follow-up.
11	Lee [53]	Comparison study	16 post-menopausal women>36%body weight	8 subjects were test group taught yoga and the other 8 were control	16 weeks	variables of body composition, visceral fat, serum adiponectin, and metabolic syndrome factors were measured in all the participants before and after the 16-week study	yoga exercise improves adiponectin level, serum lipids, and metabolic syndrome risk factors in obese postmenopausal women.
12	Nagarathna R. et al. [79]	prospective two-armed interventional randomized control study	277 type 2 diabetics of both genders aged above 28 years	study group were allocated to a yoga-based life style modification program or exercise-based life style modification program. Integrated yoga special technique for diabetes included yogasanas, pranayama, meditation and lectures on yogic life style. Control intervention included physical exercises and life style education	9 months	Medication score, blood glucose, HbA1c and lipid profile were assessed at baseline and after 9 months	Yoga based life style modification program is similar to exercise-based life style modification in reducing blood glucose, HbA1c, triglycerides, total cholesterol and VLDL. Yoga is better than exercise in decreasing oral hypoglycemic medication requirement and LDL; and increasing HDL in type 2 diabetics.
13	McDermott et al. [80]	Randomized control trial	41 participants with elevated fasting blood glucose in Bangalore, India	Study group were randomized to either yoga (n = 21) or a walking control (n = 20). Participants were asked to either attend yoga classes or complete monitored walking 3–6 days per week for eight weeks	8 weeks	Primary outcomes included changes in BMI, waist circumference, fasting blood glucose, postprandial blood glucose, insulin, insulin resistance, blood pressure, and cholesterol and also measures of psychological well-being including changes in depression, anxiety, positive and negative affect and perceived stress	Among Indians with elevated fasting blood glucose, participation in an 8-week yoga intervention resulted in greater weight loss and reduction in waist circumference when compared to a walking control. There were no between group differences in fasting blood glucose, postprandial blood glucose, insulin resistance or any other factors related to diabetes risk or psychological well-being. There were significant reductions in systolic and diastolic blood pressure, total cholesterol, anxiety, depression, negative affect and perceived stress in both the yoga intervention and walking control over the course of the study. Yoga offers a promising lifestyle intervention for decreasing weight-related type 2 diabetes risk factors and potentially increasing psychological well-being

14	V. P. Jyotsna [63]	Randomised control trial	120 patients of diabetes on oral medication along with diet and exercise	Study groups were randomized into two groups: (1) Continued to receive standard treatment for diabetes. (2) Patients administered comprehensive yogic breathing program and monitored to regularly practice yoga in addition to standard treatment of diabetes.	6 months	Glycemic control, quality of life and cardiac autonomic function tests (CAFT) were repeated after 6 months of intervention and compared with those before randomisation.	sympathetic cardiac autonomic functions significantly improved from baseline in the group practicing comprehensive yogic breathing.

Given that research is unveiling the various benefits of yoga, the application of yoga to address the various physical and mental manifestations of the "**epidemic**" diabetes looks promising and feasible. An example is that of a novel yoga program developed by a team of researchers to prevent diabetes termed YOGA-DP. Following a systematic review of trials employing yoga to address T2DM, Chattopadhyay and team (2020) selected nine randomized controlled trials (RCTs) identifying 58 yoga practices that addressed T2DM and finally 31 practices were selected to prevent T2DM in the high-risk Indian population.

Figure 7 below depicts YOGA-DP designed to prevent T2DM that is next looking at a multi-center feasibility randomized controlled trial (Copyright © 2020 Chattopadhyay, Mishra, Manjunath, Harris, Hamer, Greenfield, Wang, Singh, Lewis, Tandon, Kinra and Prabhakaran. *This is an open-access article distributed under the terms of the* ***Creative Commons Attribution License (CC BY)****. The use, distribution or reproduction in other forums is permitted, provided the original author(s) and the copyright owner(s) are credited and that the original publication in this journal is cited, in accordance with accepted academic practice. No use, distribution or reproduction is permitted which does not comply with these terms*):

Yogic practices	**Week 1–4 Each session should last for 45 min with the time split as follows:**	**Week 5+ Each session should last for 75 min with the time split as follows:**	**Details**

Shithi-likarana Vyayama	Around 5 min	Around 5 min	(1) Neck rotation *30 s* (2) Shoulder rotation *30 s* (3) Elbow flexion and extension *30 s* (4) Wrist rotation *30 s* (5) Finger movement *30 s* (6) Waist rotation *30 s* (7) Knee flexion and extension *1 min* (8) Ankle rotation *1 min* (9) Toe movement *30 s*
Surya Namaskar	–	Around 15 min	The below mentioned 12 steps constitute one set of Surya Namaskar. To complete one round of Surya Namaskar, participants need to repeat these 12 steps on the other side of their body (i.e., by extending their left leg behind in step number 4 and bringing their left leg forward in step number 9). Initially, they should practice Surya Namaskar at a slower pace. Only with practice over some time, they may try to do 12 rounds of it at a faster pace for around 15 min (i.e., a couple of seconds per step). (1) Pranamasana (prayer pose) (2) Hastauttanasana (raised arms pose) (3) Padahastasana (hands to feet pose) (4) Ashwa Sanchalanasana (equestrian pose) (5) Dandasana (stick pose) (6) Ashtanga Namaskara Asana (salute with eight parts) (7) Bhujangasana (cobra pose) (8) Parvatasana (mountain pose) (9) Ashwa Sanchalanasana (equestrian pose) (10) Padahastasana (hands to feet pose) (11) Hastauttanasana (raised arms pose) (12) Pranamasana (prayer pose)

Asana	Around 15 min	Around 30 min	Two-sided poses (right and left) are to be practiced for about 3 min (1.5 min on each side) and central-positioned poses are to be practiced for about 1.5 min. In each session, the Yogic poses are selected from the list below to prevent boredom from the similarity of routine. Advanced Yogic poses are introduced from week 5 onwards, for example, Konasana (angle pose), Trikonasana (triangle pose), Paravakonasana (lateral angle pose), Ardhaustrasana (half camel pose), Ustrasana (camel pose), Dhanurasana (bow pose), and Naukasana (boat pose). (A) Standing poses (1) Tadasana (palm tree pose) *1.5 min* (2) Ardhachakrasana (half wheel pose) *1.5 min* (3) Katichakrasana (waist wheel pose) *3 min* (4) Konasana (angle pose) or Trikonasana (triangle pose) or Paravakonasana (lateral angle pose): alternatively *3 min* (B) Sitting poses (1) Vajrasana (adamant pose) *1.5 min* (2) Mandukasana (frog pose) *1.5 min* (3) Ardhaustrasana (half camel pose) or Ustrasana (camel pose): alternatively *1.5 min* (4) Vakrasana (twisted pose) or Ardhamatsyendrasana (half spinal twist pose): alternatively *3 min* (5) Paschimottanasana (seated forward bend pose) or Janusirsasana (head to knee pose): alternatively *1.5 or 3 min, respectively* (C) Lying poses- front/prone (1) Ardhashalabhasana (half locust pose) or Poornashalabhasana (full locust pose): alternatively *3 or 1.5 min, respectively* (2) Dhanurasana (bow pose) *1.5 min* (3) Makarasana (crocodile pose) *1.5 min* (D) Lying poses- back/supine (1) Uttanapadasana (raised legs pose) or Ardhahalasana (half plow pose): alternatively *1.5 min* (2) Pavanamuktasana (wind relieving pose) *1.5 min* (3) Naukasana (boat pose) *1.5 min* (4) Saralmatsyasana (easy fish pose) *1.5 min*

Pranayama	Around 13 min	Around 13 min	(1) Vibhagiya Pranayama (sectional breathing) *4 min* (2) Nadishodhana Pranayama (alternate nostril breathing) *3 min* (3) Kapalbhati Pranayama (skull shining breathing) or Bhastrika Pranayama (bellow breathing): alternatively *3 min* (4) Bhramari Pranayama (bee breathing) *3 min*
Dhyana and relaxation practices	Around 12 min	Around 12 min	In each session, the following Dhyana and relaxation practices are to be done in a darkened room. (1) A Kara chanting, U Kara chanting, and M Kara chanting *3 min* (2) Yoga Nidra (Yogic sleep) *9 min*

2.6 REFERENCES

1. Pouya Saeedi, Inga Petersohn , Paraskevi Salpea , Belma Malanda , Suvi Karuranga , Nigel Unwin , Stephen Colagiuri, Leonor Guariguata, Ayesha A. Motala, Katherine Ogurtsova, Jonathan E. Shaw, Dominic Bright, Rhys Williams , On behalf of the IDF Diabetes Atlas Committee. Global and regional diabetes prevalence estimates for 2019 and projections for 2030 and 2045: Results from the International Diabetes Federation Diabetes Atlas, 9th edition. Diabetes research and clinical practice 157 (2019) 107843.
2. Zubin Punthakee MD, MSc, FRCPC, Ronald Goldenberg MD, FRCPC, FACE, Pamela Katz MD, FRCPC. Definition, Classification and Diagnosis of Diabetes, Prediabetes and Metabolic Syndrome Diabetes Canada Clinical Practice Guidelines Expert Committee. Can J Diabetes 42 (2018) S10–S15.
3. Martínez-Castelao A, Navarro-González JF, Górriz JL, De Alvaro F. The Concept and the Epidemiology of Diabetic Nephropathy Have Changed in Recent Years. *Journal of Clinical Medicine*. 2015; 4(6):1207-1216. https://doi.org/10.3390/jcm4061207
4. Kyeongra Yang, Lisa M. Bernardo, Susan M. Sereika, Molly B. Conroy, Judy Balk, Lora E. Burke, "Utilization of 3-Month Yoga Program for Adults at High Risk for Type 2 Diabetes: A Pilot Study", Evidence-Based Complementary and Alternative Medicine, vol. 2011, Article ID 257891, 6 pages, 2011. https://doi.org/10.1093/ecam/nep117
5. Ramamoorthi R, Gahreman D, Skinner T, Moss S (2019) The effect of yoga practice on glycemic control and other health parameters in the prediabetic state: A systematic review and meta-analysis. PLoS ONE 14(10): e0221067. https://doi.org/10.1371/journal.pone.0221067
6. Thangasami SR, Chandani AL, Thangasami S (2015) Emphasis of Yoga in the Management of Diabetes. J Diabetes Metab 6: 613. doi:10.4172/2155- 6156.1000613
7. McDermott, K.A., Rao, M.R., Nagarathna, R. *et al.* A yoga intervention for type 2 diabetes risk reduction: a pilot randomized controlled trial. *BMC Complement Altern Med* **14,** 212 (2014). https://doi.org/10.1186/1472-6882-14-212
8. Sreedevi, A., Gopalakrishnan, U. A., Karimassery Ramaiyer, S., & Kamalamma, L. (2017). A Randomized controlled trial of the effect of yoga and peer support on glycaemic outcomes in women with

type 2 diabetes mellitus: a feasibility study. *BMC complementary and alternative medicine*, *17*(1), 100. https://doi.org/10.1186/s12906-017-1574-x

9. Chattopadhyay K, Mishra P, Manjunath NK, Harris T, Hamer M, Greenfield SM, Wang H, Singh K, Lewis SA, Tandon N, Kinra S and Prabhakaran D (2020) Development of a Yoga Program for Type-2 Diabetes Prevention (YOGA-DP) Among High-Risk People in India. *Front. Public Health* 8:548674. doi: 10.3389/fpubh.2020.548674

Chapter
3 Yoga for Few Other Conditions

(Shivsanjeevi Sripathi)

CONTENTS

"Yoga is the cessation of the movements of the mind. Then there is abiding in the Seer's own form."

-Sage Patañjali

3.1 YOGA FOR CHRONIC FATIGUE:

Affecting 836,000 and 2.5 million Americans, myalgic encephalomyelitis/chronic fatigue syndrome (ME/CFS) involves an unexplained feeling of fatigue that is also associated with dysfunctioning of immunological and endocrinological aspects. The condition is attributed to a loss of $20,000 per year per house with the annual cost accounting for 17–24 billion (USD) dollars (Cortes Rivera et al, 2019). To enable the readers to get a glimpse of the affected systems, Table 1 below lists the same (Cortes Rivera et al, 2019,):

System	Tissue/Cell	Feature
Central Neurological System	Neuron	The symptomatology is related to a variety of sources of chronic neurological disturbance and associated distortions and chronicity in noxious sensory signaling and neuroimmune activation
	Glial cells	There is a significant blood–brain barrier permeability, microglia activation through toll-like receptors (TLR) signaling, secretion of IL-1B, upregulation of 5-HTT in astrocytes, reduced extracellular 5-HT levels, and hence a reduced activation of 5-HT receptors
Immune System	Lymphocytes Th1/Th2	Significant bias toward Th2 immune responses in CFS patients leading to an effector memory cell bias toward type 2 responsiveness
	NK cells	Reduction of cytotoxic activity in CFS, leading to a higher susceptibility of infection
	B cells	Persistence of autoreactive cells that can generate autoantibodies during common infections
Endocrine System	Hypothalamus–pituitary–adrenal (HPA) axis	Enhanced corticosteroid-induced negative feedback, basal hypocortisolism, attenuated diurnal variation, and a reduced responsivity to challenge

The condition is yet to see an effective approach to quell its manifestations with 85–90% cases considered undiagnosed! The various approaches are depicted in table 2 below (Cortes Rivera et al, 2019) (Cortes Rivera et al, 2019, © 2019 by the authors. Licensee MDPI, Basel, Switzerland. *This article is an open access article distributed under the terms and conditions of the Creative Commons Attribution (CC BY) license (http://creativecommons.org/licenses/by/4.0/)*:

Medication	**Examples**	**Intervention**	**Adverse Reactions**
NSAIDs	Ibuprofen, Naproxen	Relieve frequent or severe joint and muscle pain, headaches, reduce fevers and inflammation [212].	Gastrointestinal distress and bleeding
Tricyclic antidepressants	Amitriptyline, Doxepin, Nortriptyline, Desipramine	Symptom relieve, improve sleep, and relieve pain in much lower doses than those used to treat depression. Has anti-anxiety effect and improve locomotor activity [213].	Sedation, urinary retention, sexual dysfunction, weight-gain comorbidities.
Selective serotonin-reuptake inhibitors	Fluoxetine, Sertraline, Paroxetine	Helpful for anxiety/depression and other mood disorders in patients with ME/CFS, as well as patients with chronic neuropathic pain [214].	No specific adverse reactions have been described in the RCT
Antiviral Drugs	Rintatolimod, Valganciclovir	Enhance the NK-function and influence the 2-5A-synthetase pathway, producing an objective improvement in exercise tolerance and a reduction in ME/CFS-related concomitant medication usage [193,215].	Is a well-tolerated medication in the right dosage
Monoclonal Antibodies	Rituximab	Decrease the activity and number B-cell by inhibiting CD20, thus reducing inflammation. Studies demonstrate symptoms alleviation and improvement in quality of life within a 12-month follow-up [216,217].	Neutropenia, and increase of severe infections
Complementary and alternative medicine	Nutritional supplements, Acetyl-L-carnitine, Essential fatty acids, Magnesium, Vitamins, Coenzyme Q10 plus	Nutritional supplements may improve ME/CFS-related physical and mental fatigue in patients with specific nutritional deficiencies [215]. There are discrepant results in most of the RCT, and further research is needed in order to conclude a specific therapeutic role.	No specific adverse reactions have been described in the RCT of nutritional supplements
Corticosteroids	Hydrocortisone, Frudocortisone	Associated with statistical improvement in ME/CFS symptoms, especially in physical fatigue [218].	Adrenal suppression, mood disorders, weight-gain comorbidities.

This brings yoga into the scenario. Researchers Oka Tanahashi and team (2014) published a randomized controlled trial (***considered the***

first) [University Hospital Medical Information Network (UMIN CTR) UMIN000009646] in *BioPsychoSocial medicine* on the utility of yoga along with conventional medicines with individuals on pharmacotherapy alone serving as the control (15 participants each). The yoga program was isometric yoga and the system adopted is depicted in Figure 1 below: 1) Awareness of breath 2) Six poses done 4-6 times 3) Abdominal breathing 4–6) repetition (Copyright © Oka et al.; licensee BioMed Central Ltd. 2014: *This is an Open Access article distributed under the terms of the Creative Commons Attribution License (http://creativecommons.org/licenses/by/4.0), which permits unrestricted use, distribution, and reproduction in any medium, provided the original work is properly credited. The Creative Commons Public Domain Dedication waiver (http://creativecommons.org/publicdomain/zero/1.0/) applies to the data made available in this article, unless otherwise stated*):

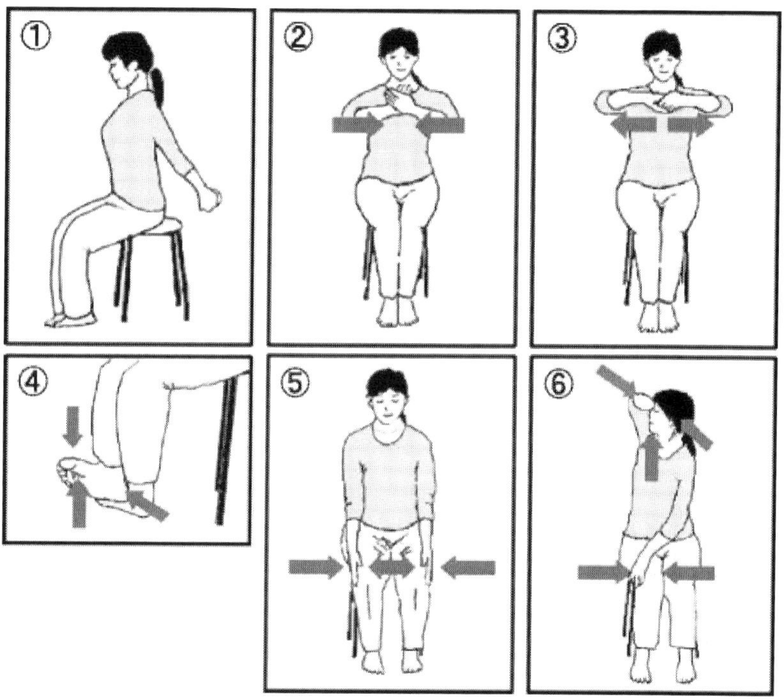

On the basis of the fatigue (F) and vigor (V) scores of the Profile of Mood States (POMS) questionnaire, the former score dipped from 21.9 ± 7.7 to 13.8 ± 6.7, $P<0.001$ while the latter score was elevated from 17.8 ± 7.6 to 22.9 ± 8.2, $P=0.002$ following the fulfillment of the yoga session with the teacher. The Chalder's fatigue scale (FS) revealed that the yoga group displayed a higher improvement in fatigue than the controls. Further, the Medical Outcomes Study Short Form 8, standard version (SF-8™) for health-related quality of life (QOL) revealed an improvement in health and lowered pain in the yoga group. To give an example of how the yoga participants felt, two patients stated "*After regular practice of yoga, I started to wake up more easily in the morning, which had been hard, because tiredness in the morning decreased.*" Further 14/15 patients found yoga to be of help and use; the one remaining patient was scared to close her eyes.

Figure 2 below is illustrative of the changes in the fatigue (F) and vigor (V) scores as per the POMS analyses; notice the lowered fatigue and increased vigor following the yoga (Copyright © Oka et al.; licensee BioMed Central Ltd. 2014: *This is an Open Access article distributed under the terms of the Creative Commons Attribution License (http://creativecommons.org/licenses/by/4.0), which permits unrestricted use, distribution, and reproduction in any medium, provided the original work is properly credited. The Creative Commons Public Domain Dedication waiver (http://creativecommons.org/publicdomain/zero/1.0/) applies to the data made available in this article, unless otherwise stated*):

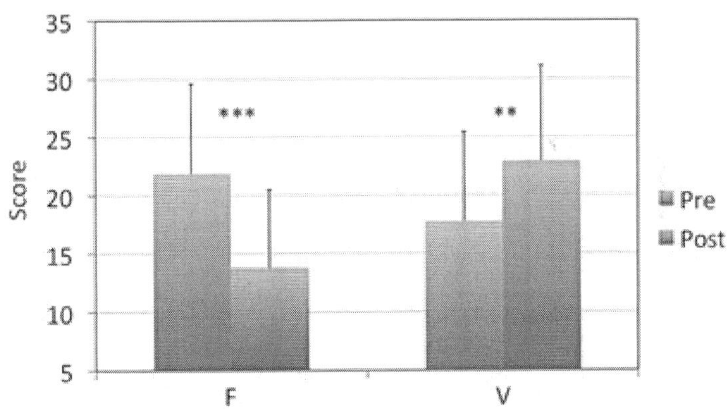

Overall, the authors showed that the results were indicative of lowering of fatigue and body pain with increased vigor and quality of life by following yoga. This becomes significant given that the participants were recruited following them not gaining much relief from pharmacotherapy/ herbal medicine/ graded exercise therapy (GET) for six months.

3.2 QUELLING PRO-INFLAMMATORY MARKERS

Given that salivary cytokines have been tested to assess the functioning of the immune system and diabetes, a research article explored the first assessment of salivary markers of inflammation (Clinical Trials.gov # NCT02108769). Randomization of twenty healthy volunteers was done into the Attention Control (AC) arm whose participants were asked to read and a Yogic Breathing (YB) arm whose participants were taught yogic breathing based on an ancient Tamil book "*Tirumanthiram*". The saliva samples were then tested for IL-1β, IL-1RA, IL-6, IL-8, IL-10, IL-17, IP-10, MCP-1, MIP-1b, and TNF-alpha: cytokines associated with stress.

While no variations were recorded for IL-1RA, IL-6, IL-10, IL-17, IP-10, MIP-1b, and TNF-alpha, the levels of IL-1β, IL-8, and MCP-1 dipped in the YB participants vs. the control. This becomes significant given that stress and inflammatory conditions are associated with activated IL-1β. Oxidative stress is associated with IL-8 whose high levels are seen in psychosomatic issues and breast cancer. Monocyte chemotactic protein (MCP-1) or chemokine (C-C motif) ligand 2 (CCL2) is aberrantly expressed in pathologies such as post-traumatic stress disorder (PTSD) and coronary disease.

Table 3 below documents changes in the three pro-inflammatory cytokines IL-1β, IL-8 and MCP-1 (pg/mL) in the control (AC) vs. yogic breathing (YB) (Twal, Wahlquist & Balasubramanian, 2016: *Open Access: This article is distributed under the terms of the Creative Commons Attribution 4.0 International License (http://creativecommons.org/licenses/by/4.0/), which permits unrestricted use, distribution, and reproduction in any medium, provided you give appropriate credit to the original author(s) and the source, provide a link to the Creative Commons license, and indicate if changes were made. The Creative Commons Public Domain Dedication waiver (http://creativecommons.org/publicdomain/zero/1.0/) applies to the data made available in this article, unless otherwise stated*):

	0 min		5 min		10 min		15 min		20 min	
	AC	YB	AC	YB	AC	YB	AC	YB	AC	YB
IL-1β (pg/mL)	432.40 (440.99)	87.04 (110.64)	231.69 (248.58)	95.43 (142.79)	184.96 (200.35)	62.96 (60.07)	149.98 (172.63)	66.12 (52.34)	215.10 (211.72)	66.18 (49.24)
IL-8 (pg/mL)	1812.13 (2304.86)	364.68 (375.22)	539.79 (336.52)	150.72 (101.32)	476.28 (275.59)	159.05 (142.59)	426.33 (283.75)	145.47 (100.20)	845.10 (915.08)	196.05 (149.71)
MCP1 (pg/mL)	329.38 (232.63)	143.03 (167.12)	201.90 (95.70)	91.21 (94.37)	186.81 (111.70)	103.95 (113.81)	149.20 (67.97)	97.47 (110.70)	275.95 (211.78)	162.32 (199.17)

(Twal, Wahlquist & Balasubramanian, 2016)

3.3 HIV

Looking at complementary approaches to augment their quality of life, a scoping review of mind-body practices was published by Ramirez-Garcia and team (2019). Among the approaches analyzed such as yoga, relaxation, Qigong and Tai chi, yoga was reported to "*show encouraging results in decreasing physical and psychological symptoms and improving quality of life and health in people living with HIV*".

Table 4 below summarizes the research studies and a snapshot of the benefits of yoga in HIV patients (Ramirez-Garcia et al, 2019: *Open Access: This article is distributed under the terms of the Creative Commons Attribution 4.0 International License (http://creativecommons.org/licenses/by/4.0/), which permits unrestricted use, distribution, and reproduction in any medium, provided you give appropriate credit to the original author(s) and the source, provide a link to the Creative Commons license, and indicate if changes were made. The Creative Commons Public Domain Dedication waiver (http://creativecommons.org/publicdomain/zero/1.0/) applies to the data made available in this article, unless otherwise stated*):

Project identifier	**Design**	**Sample size**	**Intervention/duration**	**Outcome/ time**
[35] USA	one group pretest-posttest design	48 BMSG	10 yoga, nutrition, hardiness, alternatives therapies g.s. /10 weeks	Post-int
[36] USA	one group pretest-posttest design + QA	4 YG	16 Iyengar yoga g.s. /8 weeks	Post-int ↓ pain ↓ anxiety
[37] India	one group pretest-posttest design	96 YGNA	Twice daily yoga, hydrotherapy, diet therapy and sun bath /1 to 180 days – (average- 4 weeks)	Post-int ↑ CD4 counts
[38] Canada	RCT+QA	20 YG	27 yogic breathing and meditation g.s. /14 weeks	Post-int ↑ well-being
		27 CG	usual care	
[20, 113] USA	RCT	29 YG	18 Ashtanga Vinyasa yoga i.s. +22 Ashtanga Vinyasa yoga g.s. /20 weeks	Post-int ↓ trans fat intake ↓ blood pressure
		21 CG	usual care	
[39] conference abstract India	RCT	YG 35	90 yoga s. /12 weeks	Post-int ↓ distress
		CG 35	usual care	

[40, 114] Zambia	RCT	adolescents		Post-int ↓ emotional symptom ↑ CD4 counts
		12 YGPS	20 yoga g.s. and group discussion /10 weeks	
		11 PS	20 group discussion /10 weeks	
		11 CG	usual care	
[41] India	RCT	31 YG	12 sudarshan kriya yoga g.s. of 2 h /12 weeks	Post-int ↑ QoL
		30 CG	usual care	
[42] USA	pilot RCT	12 YG	16 yoga g.s. /8 weeks	Post-int & 4 months Feasible and acceptable ↓ distress ↓ stress

3.4 FINAL POINTS

Before wrapping up, a meta-analysis exploring the effects of yoga on the quality of life in cancer survivors between January 1970 and July 2010 reported evident improvements in terms of stress, distress, anxiety and depression vs. controls. Table 5 below summarizes the studies and effects as outlined by the researchers Kuan-Yin Lin et al, 2010 ():

Abbreviations: *QOL: quality of life; HADS: Hospital Anxiety and Depression Scale; PSS: Perceived Stress Scale; STAI: state trait anxiety inventory; SF-12: The 12-Item Short Form Health Survey; FACT-B: Functional Assessment of Cancer Therapy-Breast; FACT-Fatigue: Functional Assessment of Cancer Therapy-Fatigue; FACIT-Sp: Functional Assessment of Chronic Illness Therapy-Spirituality; CES-D: Center for Epidemiologic Studies Depression Scale; PSQI: Pittsburgh sleep quality inventory; PANSA: positive and negative affect schedule; FACT-G: The Functional Assessment of Cancer Therapy-General; FACIT-Fatigue: Functional Assessment of Chronic Illness Therapy-Fatigue; POMS: profile of mood states; SOSI: symptoms of stress inventory; EORTC QLQ-C30: European Organization for Research and Treatment of Cancer Quality of Life Questionnaire Version 3.0; LSI: The Leisure Score Index; IES: Impact of Events Scale; STATE:*

Speilberger State Anxiety Inventory; BFI: Brief Fatigue Inventory; NA: not available; MBSR: mindfulness-based stress reduction; SF: Medical Outcomes Studies Short-Form General Health Survey; MOSS: Medical Outcomes Social Support Survey; MBAT: mindfulness-based art therapy; SCL-90-R: Symptoms Checklist Revised; SF-36: Medical Outcomes Study Short-Form Health Survey; CT: chemotherapy; min: minute; hr: hour.

Author, year	No. of participants	Age, mean ± SD	Yoga type	Frequency	Duration, week	Outcome measures	Results
Danhauer et al., 2009 [17]	44	55.8 ± 9.9	Restorative yoga: asanas (postures done with awareness), pranayama (voluntarily regulated nostril breathing), and savasana (deep relaxation)	75 min weekly	10	QOL (SF-12 and FACT_B), fatigue (FACT-Fatigue), depression (CES-D), positive and negative affect (PANAS), and spirituality (FACIT-Sp), sleep quality (PSQI)	Group differences favoring yoga group in mental health, depression, positive affect, and spirituality. Significant baseline*group interactions in negative affect and emotional well-being.
Raghavendra et al., 2009 [18]	88	46.0 ± 9.1	Integrated yoga: asanas, pranayama, meditation, and yogic relaxation	1 hr, 3 sessions weekly	6	Anxiety and depression (HADS), level of stress (PSS)	Significant between group differences on anxiety ($P < .001$), depression ($P = .002$), perceived stress ($P < .001$), salivary cortisol ($P = .009$), and pooled mean cortisol ($P = .03$). significant positive correlation between morning salivary cortisol level and anxiety and depression
Rao et al., 2009 [19]	98	NA	Integrated yoga: asanas, pranayama, meditation, and yogic relaxation	60 min daily	24	Anxiety (STAI)	Overall decrease in both self-reported state anxiety ($P < .001$) and trait anxiety ($P = .005$)
Lengacher et al., 2009 [20]	84	57.5 ± 9.4	MBSR: gentle yoga	2 hr weekly	6	Fear of recurrence (Concerns about Recurrence Scale, anxiety (STAI), depressive symptoms (CES-D), optimism (Life Orientation Test), perceived stress (PSS), QOL (SF), social support (MOSS), spirituality	Significant between group differences on levels of depression, anxiety, fear of recurrence, energy, physical functioning, and physical role functioning (two-sided $P < .05$)

Banerjee et al., 2007 [21]	68	44.0 ± 1.4	Integrated yoga: asanas, pranayama, nidra (guided relaxation with imagery)	90 min weekly	6	Anxiety and depression (HADS), level of stress (PSS), DNA damage	Significant between group differences on HADS score, mean PSS, and DNA damage ($P < .001$)
Moadel et al., 2007 [22]	128	54.8 ± 9.9	Hatha yoga: physical stretches, breathing, and meditation	1.5 hr weekly	12	QOL (FACT-G), spiritual well-being (FACIT-Sp), fatigue (FACIT-Fatigue), distress (Distressed Mood Index)	Significant between group differences on social well-being ($P < .0001$). For patients not receiving CT during intervention, significant between group differences on overall QOL, emotional well-being, social well-being, spiritual well-being, and distress mood
Culos-reed et al., 2006 [23]	38	51.1 ± 10.3	Yoga: asanas, shevasana (relaxation)	75 min	7	Anxiety and depression (POMS), QOL (EORTC QLQ-C30), symptom of stress (SOSI), physical activity (LSI), fitness (grip strength, flexibility, and Rockport Walking Test)	Significant differences between groups at post-intervention in global QOL, emotional function, diarrhea, and tension ($P < .05$).
Monti et al., 2006 [24]	111	53.6 ± 11.5	MBAT: gentle yoga	2.5 hr weekly	8	Psychological distress and stress-related somatic complaints (SCL-90-R), Health-related QOL (SF-36)	Significant between group differences on symptoms of distress ($P < .001$), mental composite of SF-36 ($P = .025$), general health ($P = .008$), mental health ($P < .001$), social functioning ($P = .048$), and vitality ($P = .01$)
Cohen et al., 2004 [25]	39	51	Tibetan yoga: controlled breathing, mindfulness, postures from Tsa lung (channels and vital breath), Trul khor (magical wheel)	weekly	7	Distress (IES), anxiety (STATE), depression (CES-D), fatigue (BFI), sleep disturbances (PSQI)	Significant between group differences on sleep disturbance scores ($P < .004$), sleep quality ($P < .02$), sleep latency ($P < .01$), sleep duration ($P < .03$), use of sleep medications ($P < .02$). no significant differences between groups in terms of intrusion, state anxiety, depression, or fatigue

Speca et al., 2000 [26]	90	51	MBSR: gentle yoga	90 min weekly	7	Anxiety and depression (POMS), stress-related symptoms (SOSI)	Significant between group differences on total mood disturbance, subscales of depression, anxiety, anger, confusion, vigor, and symptoms of stress.

Overall, if patients are considered, Arndt Büssing and team (2012) offer a snapshot of the benefits of yoga to help ameliorate their ailments and psyche in table 6 below (Copyright © 2012 Arndt Büssing et al. *This is an open access article distributed under the Creative Commons Attribution License, which permits unrestricted use, distribution, and reproduction in any medium, provided the original work is properly cited*):

	Specific effects	**Unspecific effects**
Cognition	Contemplative states; Mindfulness; Self-identity; Self-efficacy; Beliefs; Expectations	Control of attentional networks
Emotions	Emotional control/regulation	Quality of Life
Physiology	Vagal afferent activity; Heart rate/ Respiratory; Relaxation response/ Stress reduction	Social contacts
Physical body	Physical flexibility, Fitness/Endurance	Healthy life style
Specific and unspecific effects are often interconnected.		

Further and continued research involving more trials can propel the use of the ancient approach across the world to quell diseases and usher in hope, health and harmony!

3.5 REFERENCES

1. Cortes Rivera, M., Mastronardi, C., Silva-Aldana, C. T., Arcos-Burgos, M., & Lidbury, B. A. (2019). Myalgic Encephalomyelitis/Chronic Fatigue Syndrome: A Comprehensive Review. *Diagnostics (Basel, Switzerland)*, *9*(3), 91. https://doi.org/10.3390/diagnostics9030091
2. Oka, T., Tanahashi, T., Chijiwa, T., Lkhagvasuren, B., Sudo, N., & Oka, K. (2014). Isometric yoga improves the fatigue and pain of patients with chronic fatigue syndrome who are resistant to conventional therapy: a randomized, controlled trial. *BioPsychoSocial medicine*, *8*(1), 27. https://doi.org/10.1186/s13030-014-0027-8
3. Twal, W.O., Wahlquist, A.E. & Balasubramanian, S. Yogic breathing when compared to attention control reduces the levels of pro-inflammatory biomarkers in saliva: a pilot randomized controlled trial. *BMC Complement Altern Med* 16, 294 (2016). https://doi.org/10.1186/s12906-016-1286-7
4. Ramirez-Garcia, M., Gagnon, MP., Colson, S. *et al.* Mind-body practices for people living with HIV: a systematic scoping review. *BMC Complement Altern Med* **19,** 125 (2019). https://doi.org/10.1186/s12906-019-2502-z.
5. Lin, K. Y., Hu, Y. T., Chang, K. J., Lin, H. F., & Tsauo, J. Y. (2011). Effects of yoga on psychological health, quality of life, and physical health of patients with cancer: a meta-analysis. *Evidence-based complementary and alternative medicine : eCAM*, *2011*, 659876. https://doi.org/10.1155/2011/659876
6. Arndt Büssing, Andreas Michalsen, Sat Bir S. Khalsa, Shirley Telles, Karen J. Sherman, "Effects of Yoga on Mental and Physical Health: A Short Summary of Reviews", Evidence-Based Complementary and Alternative Medicine, vol. 2012, Article ID 165410, 7 pages, 2012. https://doi.org/10.1155/2012/165410

Chapter 4

Yoga an Effective Strategy for Self-Management of Stress-Related Problems and Wellbeing during COVID19 Lockdown: A Cross-Sectional Study

Pooja Swami Sahni[1], Kamlesh Singh[1,2], Nitesh Sharma[1], Rahul Garg[1,3,4]

[1] National Resource Centre for Value Education in Engineering, Indian Institute of Technology Delhi, Noida, India,

[2] Department of Humanities and Social Sciences, Indian Institute of Technology Delhi, New Delhi, India,

[3] Department of Computer Science and Engineering, Indian Institute of Technology Delhi, New Delhi, India

[4] Amar Nath and Shahsi Khosla School of Information Technology, Indian Institute of Technology Delhi, New Delhi, India

ABSTRACT

This cross-sectional research aims to study the effect of yoga practice on the illness perception, and wellbeing of healthy adults during 4–10 weeks of

---

Citation: Sahni PS, Singh K, Sharma N, Garg R (2021) Yoga an effective strategy for self-management of stress-related problems and wellbeing during COVID19 lockdown: A cross-sectional study. PLoS ONE 16(2): e0245214. https://doi.org/10.1371/journal.pone.0245214

lockdown due to COVID19 outbreak. A total of 668 adults (64.7% males, M = 28.12 years, SD = 9.09 years) participated in the online survey. The participants were grouped as; yoga practitioners, other spiritual practitioners, and non-practitioners based on their responses to daily practices that they follow. Yoga practitioners were further examined based on the duration of practice as; long-term, mid-term and beginners. Multivariate analysis indicates that yoga practitioners had significantly lower depression, anxiety, & stress (DASS), and higher general wellbeing (SWGB) as well as higher peace of mind (POMS) than the other two groups. The results further revealed that the yoga practitioners significantly differed in the perception of personal control, illness concern and emotional impact of COVID19. However, there was no significant difference found for the measure of resilience (BRS) in this study. Yoga practitioners also significantly differed in the cognitive reappraisal strategy for regulating their emotions than the other two groups. Interestingly, it was found that beginners -those who had started practicing yoga only during the lockdown period reported no significant difference for general wellbeing and peace of mind when compared to the mid- term practitioner. Evidence supports that yoga was found as an effective self-management strategy to cope with stress, anxiety and depression, and maintain wellbeing during COVID19 lockdown.

INTRODUCTION

A report by the World Economic Forum estimates that about 2.6 billion people around the world have been in some kind of lockdown that may lead to second form of stress-related disorder epidemic in the second half of 2020 [1]. Similar to the World economic forum estimations, a survey by the Indian Psychiatric Society shows that two-fifth of the people are experiencing common mental disorders, due to lockdown and the prevailing COVID19 pandemic in India [2]. This indicates the need for an urgent action to reduce the adverse effects of the COVID19 lockdown on the general well-being of people.

Various factors have been suggested to be contributing to the worsening of mental health. One of the major factors reported causing stress, anxiety and depression is fear of getting infected with the virus/disease (COVID19) [3]. Previous studies examining illness perception in the context of other chronic diseases such as; diabetes, AIDS and myocardial infarction found that people create their representations of the illness related to its risk of contracting, cause, time the illness will last for, and the consequence of the illness [4–6]. Further, it is also suggested that beliefs plays an important role in the way people create notions for the controllability and cure of the illness [7]. These representations and notions are argued to determine the stress response, and the ways of coping, which is believed to affect the wellbeing [8]. On the other hand, some of the factors that influences the perceived effects of the illness are suggested to be the physical and mental health status of the individuals, with healthy people reporting less cognitive and emotional representations of illness [9]. Even the type of treatment and the preventive measures received are suggested to drive the illness coherence/ understanding and perception about personal control over the illness [10].

Even though there are some studies supporting that yoga can be used as complementary and alternate therapy for mental health, there is need for empirical research studies to provide evidence for yoga as effective strategy for self-management of stress-related problems during COVID19. Further, to the best of our knowledge the empirical investigation for the effects of yoga and other spiritual practices on illness perception and wellbeing related problems experienced by people during COVID19 has not been examined so far. The present research uses a cross-sectional study design to examine the effect of the practice of yoga and other spiritual practices on illness perception, and wellbeing of adults. In this study, wellbeing has been assessed through measures of depression, stress, anxiety, resilience, peace of mind and the strategies employed to regulate the emotional upheavals. This approach has been reported in earlier studies that have examined wellbeing in terms of anxiety, stress, and depression [11–13], emotion regulation [14] and as a measure of peace of mind [15]. Wellbeing has also been shown to positively correlate with resilience [16]

Previous research has suggested that yoga can be used as a non-pharmaceutical measure or as a complement to drug therapy for treatment or cure of modern epidemic diseases like mental stress, obesity, diabetes, hypertension, coronary heart disease, and chronic obstructive pulmonary disease [17]. Some recent studies also propose that yoga can assume a ground-breaking complementary and alternative therapy in the battle against the novel coronavirus while improving the physical and mental wellbeing of people in this pandemic circumstance [18, 19]. However, for long researchers have debated the role of yoga and other spiritual practices in the mental health of people [20]. The practice of yoga is most commonly perceived as physical exercise that helps in gaining flexibility, physical strength and helps to relax. In Indian philosophical texts, yoga is treated as a spiritual practice that is related to training of the mind. Patanjali yoga sutra describes yoga as a practice of *'Chitta vritti nirodha'* literally translating as controlling or calming of the mind. Most commonly, a typical yoga schedule follows a combination of *asanas* (postures), *pranayama* (breath control), *pratyahara* (withdrawal of senses), *Dharana* (concentration), *dhyana* (meditation) and *Samadhi* (absorption). While *asanas* are reported to help in improving the physical strength and flexibility, it is argued to also help in building concentration [21]. Preliminary research suggests that *pranayama* calms the nervous system and helps in regulating the blood pressure [22]which is further argued to improve the stress response. *Pratyahara, Dharana, Dhyana* includes techniques such as; *mantra* chanting, *yoga nidra*, and *antar mouna* that are said to help in developing an ability to internally witness the sensory inputs [23]. This witnessing capacity is speculated to help one in the reappraisal of the problem in hand, control the fluctuations of the mind and reduce the unconscious negative mental perceptions. Apart from yoga, there are some other spiritual practices such as listening to *satsang* [24], *swadhyaya* (reading Holy Scriptures) [25] and rendering *seva* (selfless service to the community) that have been reported to help maintain wellbeing, reduce stress, anxiety and depression [26, 27].

MATERIALS AND METHODS

In this research three groups; yoga practitioners, other spiritual practices and non-practitioners were examined for the differences in the measures of illness perception and wellbeing. Additionally, differences based on the duration of practice were also examined in three categories; long-term, mid- term and beginner group. A cross-sectional study was designed using standardized scales for wellbeing related measures and illness perception, questions about daily practices and demographics. An online questionnaire booklet was prepared using google forms and data was collected via social networking groups and email mailing lists. The responses were analyzed using SPSS ver. 26.0.

Participants

The sample consisted of a total of 668 adults, out of which 96.26% (n = 643) chose to participate and 3.74% (n = 25) declined. Out of the 643 participants, there were 64.7% males (n = 416), 34.7%, females (n = 223), and four preferred not to say about their gender. The age range was 18–72 years (M = 28.12 years, SD = 9.09 years). In the sample, a total of 59% (n = 380) were students and 41% (n = 263) were from non- student groups. The non-student group had 34.4% (n = 221) working adults and the remaining 6.5% (n = 42) were from other categories (retired, homemaker). Concerning qualification, 53.2% (n = 342) participants were from postgraduate and higher qualification, 20.8% (n = 134) from graduate, 23.3% (n = 150) from intermediate or pursuing graduate, and the remaining 2.6% (n = 17) were from high school or below education level. Majority (73.2%) of the participants were from urban (n = 471), whereas 16.5% (n = 106) were from semi-urban, and 10.26% (n = 66) reported being from rural areas of residence.

Within the study sample, 59.6% (n = 384) reported that they practice yoga (includes asana, pranayama, meditation, mantra chanting or any other form of yogic practice) and were categorized as yoga practitioners and 40.4% (259) responded not following any yogic practice. Out of the participants that reported not following any yogic practice, 17.6% (n = 113) reported following one or more of the other forms of spiritual practices for example; watching online spiritual programs (50%), online satsang (14.37%), reading holy scriptures(23.25%), performing seva (12.5%), and were categorized as other spiritual practitioner group.

The remaining 22.7% (n = 146) participants reported that they do not follow yogic or any other spiritual practices and they were termed as a non-practitioner group. Further, among the yoga practitioners, 35% (n = 134) were beginners (those who had started yoga practice during COVID19 lockdown period), 39.7% (n = 152) were mid- term (1≤year of practice≤4), and 25.32% (n = 97) were Long term (≥ 5 years of practice) practitioners. Within the beginners, 39.9% reported practicing yoga for all days in the week, 23.9% for 5–6 days, 23.2% for 3–4 days and 13% for 1–2 days in a week. For mid-term practitioners 32.4% reported practicing yoga for all days in the week, 30.4% for 5–6 days, 29.7% for 3–4 days and 7.4% for 1–2 days in a week.

For long term practitioners, 58.8% reported practicing yoga for all days in the week, 19.6% for 5–6 days, 11.3% for 3–4 days and 10.3% for 1–2 days in a week. The data for frequency in terms of hours of practice per day was asked as an open choice question,' How many minutes in a day do you practice yoga? Beginners reported on an average spending 31.24 mins, mid-term practitioner average of 39.10mins, and long term practitioners reported spending average of 51.25 mins, for their daily yoga practice.

Materials

Brief Illness Perception

(BIPQ) [3] was used to measure the individual's perception of COVID19 by adapting the Brief Illness Perception Questionnaire. The adapted version of BIPQ had 12 items designed to rapidly assess the cognitive and emotional representations of COVID19 illness. The five dimensions of cognitive representation of COVID19 illness was assessed through 9 items; identity—the label the person uses to describe the illness and the symptoms they view as being part of the disease (sample item: How much do you think the infected person experiences symptoms from this illness?); consequences—the expected effects and outcome of the illness (sample item: How much does this illness affect the person who suffers from it?); understanding—personal understanding about the cause of the illness (sample item: How well do you feel you understand this illness?); timeline—how long the patient believes the illness will last (sample item: How long do you think the illness last for those who have it?); and cure or control—the extent to which the patient believes that they can recover from or control the illness (sample item: How much control do you feel you have over this illness?). The emotional representation of COVID19 was assessed by 2 items incorporating negative reactions such as fear, anger, and distress (sample item: How much does thinking about this illness affect you emotionally? e.g. does it make you angry, scared, upset or depressed?). Assessment of the causal representation is by an open-ended response item adapted from the IPQ-R, which asks patients to list the three most important causal factors in rank order. All of the items except the causal question are rated using a 0-to10 response scale. The higher the score is, the greater the perception of the illness for that particular item. The total scale alpha coefficient in this study sample was 0.64.

Depression, Anxiety and Stress Scale

(DASS-9) [28] was used to measure the depression, anxiety and stress experienced by the participants during the COVID19 lockdown period using DASS-9. It is the shorter version of DASS-42 [29] and consists of three sub-factors with 3 items each viz., depression (sample item: I found it difficult to work up the initiative to do things), anxiety (sample item: I experienced trembling eg. in the hands), and stress (sample item: I tended to overreact to). The instructions were modified to suit the current research and participants were asked to rate how much each statement applied to them during the lockdown period. Cronbach's alpha for the total DASS-9 was reported by Yusoff (2013) equal to .72 whereas for Depression, Anxiety and Stress factors, alphas were .52, .57, and .55, respectively [29]. In this study sample the Cronbach alpha for total DASS-9, depression, anxiety and stress was found to be 0.73, 0.63, 0.64, and 0.53 respectively.

Scale of General Wellbeing

(SGWB) [30] was used to measure the general wellbeing through fourteen common constructs as indicators of well-being viz., happiness, vitality, calmness, optimism, involvement, self-awareness, self-acceptance, self-worth, competence, development, purpose, significance, self-congruence and connection (sample item: I accept most aspects of myself). All items were phrased positively and rated on a Likert scale from 1 (Not at all) to 5 (very true), indicating experiences in life overall. Previous studies have reported a Cronbach alpha for SGWB ranging from 0.82 to 0.92 [30]. In this study sample, the Cronbach alpha for the total scale was found to be 0.93.

Brief Resilience Scale

(BRS) [31] was used to measure resilience. The scale contains 6 items measuring the ability to bounce back from stress and difficulties (e.g., "I usually come through difficult times with little trouble"). The items are rated on a 5-point Likert scale from 1 (Strongly Disagree) to 5 (Strongly Agree). The possible score ranges from 1 (minimum resilience) to 6 (maximum resilience). Three items are negatively worded and are reversed scored. Adequate reliability, with Cronbach's alpha ranging from .80 to .91 in 4 different samples was reported in an earlier study [31]. In this study sample, the Cronbach alpha was found to be 0.73.

Peace of Mind Scale

(POMS) [15] was used to measure peace of mind through a single factor model presented by POMS. The scale consists of a 5 item scale that measures affective wellbeing. The items reflect the experiences of internal peace and harmony in general terms (e.g., I have peace and harmony in my mind) as well as in everyday circumstances (e.g., I feel content and comfortable with myself in daily life). Participants indicated how often they experience the internal states described in each of the items on a scale of 1 (not at all) to 5 (all the time). The five-item POMS (Cronbach alpha = 0.78) was used in this study which had previously been confirmed for the Indian population [32]. In this study sample, the Cronbach alpha was found to be 0.91.

Emotion Regulation Questionnaire

(ERQ) [33] was used to assess the commonly used strategies to alter emotion through ERQ viz., 6 items on cognitive reappraisal (sample items: "When I'm faced with a stressful situation, I make myself think about it in a way that helps me stay calm"), and 4 items on expressive suppression (sample items: "When I am feeling negative emotions, I make sure not to express them"). Participants responded to each item using a 7-point Likert scale ranging from 1 (strongly disagree) to 7 (strongly agree). The average of all the scores in each subscale of cognitive reappraisal and expressive suppression are used for analysis. The higher the score represents greater the use of that particular emotion regulation strategy, conversely lower scores means less frequent use. In a four study sample reported the Cronbach alpha ranging for reappraisal facet 0.75 to 0.82 and suppression factor 0.68–0.76 [34]. In this study sample, the Cronbach alpha for cognitive reappraisal and expressive suppression was found to be 0.83 and 0.75 respectively.

Apart from the standardized scales as described above, the data collection booklet included consent form, yoga schedule, and demographic information schedule. The categorization of the yoga practitioners, and non-practitioners was based on the dichotomous question; 'Do you practice yoga (includes asana, pranayama, meditation, mantra chanting or any other form of yogic practice) in your daily routine?" The non-yoga practitioner group was further classified based on multiple response question; 'Any other form of spiritual practice do you follow?' for example; online *Satsang* (listening to devotional songs), watching spiritual programs, reading Holy Scriptures, selfless service or any other. This formed a group of other spiritual practitioners and the rest of non-yoga practitioners formed a third group classified as non-practitioners. The yoga practitioners were also asked about the duration of their practice. The demographic profile consisted of information about age, gender, qualification, working status and place of residence. An additional item on the three most important causal factors (in rank order) of stress during lockdown was also asked through a question 'Please list in rank-order the THREE most important factors that you believe are reasons for stress due to lockdown'.

Procedure

Preparation for the Study

The study was designed for both Hindi and English speaking population keeping in mind the diversity in the preference for language in the population. At the outset, the original English scales for which the Hindi version was not available (ERQ and BRS), were translated into Hindi by a bilingual expert working in the area of psychological assessment. The Hindi translations of all the scales were then evaluated by the first and second author to check for adequacy of translation. Modifications were made wherever the Hindi translations were not found to adequately capture the intended meaning. Further, a bilingual expert independently back-translated these scales from Hindi to English. The back translations were again reviewed by the first and second author and matched to the original scales. At this stage, most items were found to aptly represent the content of the original English scales. The finalized Hindi and English scales were used to prepare the data collection booklet.

Data Collection and Analysis

The cross-sectional study was conducted using an online survey. The sample for the study was recruited through the distribution of an online message consisting a brief introduction to the study and a link to the google form of data collection booklet using social networks, mailing lists and snowballing techniques. The online message with the link was especially circulated among the yoga practitioner groups. An electronic consent was obtained from each of the participants before beginning the survey. The data collected was anonymous and no personal details that could identify the participants were asked in the google form. Participants were assured that the data will be kept confidential and only be used for research purposes. The google form was available for responses from 26th April -8th June 2020 (beginning of Unlock 1.0), corresponding to four to ten weeks of the lockdown in India.

In the first step of data analysis, the responses for the Hindi ($n = 96$) and English ($n = 547$) scale were analyzed using an independent t-test that showed no significant difference in illness perception or wellbeing related measures in any of the practitioner groups due to difference in language (all $ps>0.05$). Therefore, the remaining analysis was conducted on the combined English and Hindi data. In the next step, the descriptive analysis was conducted and the internal reliability scores (Cronbach alpha) for each scale was computed.

To confirm the factor structure of the scales used for this study sample, the data were subjected to confirmatory factor analysis. Multivariate analysis (MANOVA) was conducted to examine the differences in the mean scores of illness perception and wellbeing related measures among the yoga practitioners, other spiritual practitioners, and non- practitioner groups. The open-ended question was analyzed using percentage analysis. Lastly, MANOVA was also performed within the yoga practitioner group based on the duration of practice.

RESULTS

Descriptive Analysis

The descriptive statistics of all the dependent variables were analyzed. Three outliers identified based on extreme values more than three IQR's (interquartile range) [35] computed from Tukey's hinges in SPSS, were deleted. Confirmatory factor analysis (CFA) showed that most of the fit statistics for all the scales were in the acceptable range. Statistics presented in Table 1.

Table 1. Goodness of fit statistics for confirmatory factor analysis.

Measures of goodness fit	Acceptable level	BIPQ	DASS	POMS	SGWB	BRS	ERQ
$\chi2$ (df)		118.75 (26)	86.78 (24)	16.66 (3)	281.53 (70)	32.09 (5)	202.38 (34)
$\chi2$ /df	<5 [36]	4.57	3.61	5.55	4.02	6.41	5.95
p-value	>0.05	0.000	0.000	0.001	0.000	0.000	0.000
RMSEA	<0.10 [37]	0.077	0.065	0.084	0.062	0.093	0.092
CFI	≥0.95 [38]	0.915	0.953	0.995	0.967	0.970	0.927
GFI	>0.90 [39]	0.961	0.971	0.990	0.941	0.982	0.941
AGFI	>0.90 [39]	0.932	0.945	0.949	0.912	0.926	0.904

Demographic Variables

The relationship between demographic variables (age, gender, qualification, working status, and place of residence) was examined on illness perception (BIPQ), wellbeing related measures (DASS, SGWB, POMS, BRS) and Emotion regulation strategies (ERQ). MANOVA results indicated no statistically significant main effect of working status and qualification on the COVID19 perception, wellbeing measures (DASS, SGWB, POMS, and BRS) or emotion regulation strategies (ERQ). However, there was a significant effect of gender on the mean scores of illness concern F (1,634) = 11.14, $p<0.001$, partial eta squared = .02, and emotional representation of COVID19 F (1,634) = 10.50, $p<0.001$, partial eta squared = .02, partial eta squared = .02, with females reporting higher illness concern (M = 8.18, SD = 2.20),and higher emotional impact of COVID19 (M = 5.73, SD = 3.03) than males illness concern (M = 7.50, SD = 2.83), and emotional impact of COVID19 (M = 5.00, SD = 3.03), respectively. There was a significant effect place of residence on illness consequence of COVID19 F (1,634) = 5.61, $p<0.05$, partial eta squared = -.01, with urban participants reporting higher concern for consequences of COVID19 (M = 7.76, SD = 5.34) than semi-urban (M = 6.95, SD = 2.55) or rural (M = 7.22, SD = 2.88).

Age had a significant effect on depression (DASS-D) F (1,634) = 9.34, $p<0.005$, partial eta squared = -.01 and Peace of Mind (POMS) F (1,634) = 13.02, $p<0.001$, partial eta squared = -.02, with participants from age group 18–25 years reporting higher depression (DASS-D)(M = 0.97, SD = 0.70) than age group 26–35 years (M = 0.81, SD = .60), age group 36–45 years (M = 0.64, SD = 0.62) and age group 46 and above (M = 0.61, SD = 0.57). Lower mean scores for Peace of Mind (POMS) were reported by the participants of age group 18–25 years (M = 3.09, SD = 0.94) than age group 26–35 years (M = 3.42, SD = 1.07), age group 36–45 years (M = 3.55, SD = 1.07) and age group 46 and above (M = 3.80, SD = 1.00).

Effect of Yoga and Other Spiritual Practice on Illness Perception, And Wellbeing Measures

Before conducting the MANOVA, Pearson correlation was performed between all dependent variables to test the multivariate assumption that the dependent variables would be correlated with each other in the moderate range [40]. A meaningful pattern of correlations was observed amongst most of the dependent variables (r = - 0.460 to r = 0.448), suggesting the appropriateness of MANOVA. Correlations are presented in S1 Table. Additionally, the BOX's M value of 575.82 was associated with less than a p-value of 0.001, which was interpreted as non-significant based on Huberty and Petoskey's (2000) guidelines (i.e $p < .005$) [41]. Thus, the covariance matrices between the groups were assumed to be equal for the MANOVA.

The MANOVA was conducted to test the hypothesis that there would be one or more mean differences between the spiritual practitioner levels (yoga practitioners, other spiritual practitioners, and non-practitioners) and COVID19 perception, wellbeing related measures, as well as in their emotion regulation strategies. After controlling for the confounding effect of demographic variables a statistically significant MANOVA effect was obtained, Pillai's' Trace = .19, F (38, 1220) = 3.13, $p<0.001$. The multivariate effect size was estimated at .156. A series of Levene's *F* tests, to examine the homogeneity of variance assumption was conducted and the statistics are presented in Table 2. In the third stage, a series of post hoc analyses (Tukey's HSD) were performed to examine the individual mean difference comparisons across all three groups. The summary of post hoc comparisons is presented in Table 3.

Table 2. One way ANOVA with illness perception (IP), wellbeing measures, and emotion regulation (ERQ) as dependent variable and practitioner groups as the independent variable.

	Levene's statistics		**ANOVA**		
	F	**Sig.**	**F**	**Sig.**	**Partial Eta Squared**
Consequence(IP1)	2.440	.088	.677	.508	.002
Timeline (IP2)	3.594	.028	1.349	.260	.004
Personal control (IP3)	1.377	.253	16.971	.000	.051
Treatment Control (IP4)	.126	.881	2.877	.057	.009
Identity (IP5)	1.216	.297	.850	.428	.003
Illness concern(IP6)	2.831	.060	1.166	.312	.004
Coherence/understanding (IP7)	.844	.431	5.884	.003	.018
Emotional representation (IP8)	4.173	.016	4.413	.012	.014
Risk perception (IP10)	.598	.550	3.724	.025	.012
Risk perception (IP11)	.021	.980	4.460	.012	.014
Personal preventative control(IP12)	.976	.378	7.532	.001	.023
Depression (DASS-D)	6.441	.002	19.463	.000	.058
Anxiety (DASS-A)	.136	.873	.355	.701	.001
Stress (DASS-S)	1.103	.333	4.955	.007	.015
Peace of Mind (POMS)	.129	.879	40.851	.000	.114
Wellbeing (SGWB)	2.880	.057	40.271	.000	.112
Resilience(BRS)	.548	.578	1.696	.184	.005
Emotion regulation-Cognitive Appraisal (ERQ-C)	.731	.482	19.970	.000	.059
Emotion regulation-Expressive Suppression (ERQ-E)	.174	.841	.419	.658	.001

Table 3. Summary of the post hoc comparison for the three groups- yoga practitioners, other spiritual practitioner and non- practitioner groups.

Measures	Yoga Vs other Spiritual practitioner group	Other spiritual vs non-practitioner group	Yoga vs non-practitioner group
Consequence(IP1)	n.s.	n.s.	n.s.
Timeline (IP2)	n.s.	n.s.	n.s.
Personal control (IP3)	p<0.05	n.s.	p<0.05
Treatment Control (IP4)	n.s.	n.s.	n.s.
Identity (IP5)	n.s.	n.s.	n.s.
illness concern(IP6)	n.s.	n.s.	n.s.
coherence/understanding (IP7)	p<0.05	n.s.	p<0.05
Emotional representation (IP8)	p<0.05	n.s.	p<0.05
Risk perception (IP10)	p<0.05	n.s.	p<0.05
Peer Risk perception (IP11)	p<0.05	n.s.	p<0.05
Personal preventative control(IP12)	p<0.001	n.s.	p<0.001
Depression (DASS-D)	p<0.001	n.s.	p<0.001
Anxiety (DASS-A)	p<0.05	n.s.	p<0.05
Stress (DASS-S)	p<0.001	p<0.05	p<0.001
Peace of Mind (POMS)	p<0.05	p<0.05	p<0.05
Wellbeing (SGWB)	p<0.05	n.s.	p<0.05
Resilience(BRS)	n.s.	n.s.	n.s.
Emotional regulation-Cognitive Appraisal (ERQ-C)	p<0.05	p<0.05	p<0.05
Emotional regulation-(Expressive Suppression (ERQ-E)	n.s.	n.s.	n.s.

COVID19 Perception-BIPQ

There was a statistically significant difference in group means for the personal control (IP3) $F(2,639) = 14.81, p < .001$, Coherence/ understanding (IP7) $F(2,639) = 4.95, p < .01$ Emotional representation (IP8) $F(2,639) = 5.17, p < .0.01$, Risk perception (IP10 and IP11) $F(2,639) = 4.01, p < .01$, Personal preventive Control (IP12) $F(2,639) = 6.30, p < .01$. However, there were no significant differences found in the COVID19 representation of illness perception with respect to consequence (IP1), timeline (IP2), treatment control (IP4), identity (IP5) and illness concern (IP6).

A post hoc analysis (Tukey HSD) revealed significant differences in mean scores of yoga practitioner group for personal control (IP3)(M = 6.22, SD = 2.72), coherence/understanding (IP7) (M = 7.58, SD = 2.22), and Emotional representation (IP8) (M = 4.95, SD = 3.20) when compared to the mean scores of other spiritual practitioner group for personal control (M = 5.06, SD = 2.62), coherence/understanding (IP7)(M = 7.29, SD = 2.44), emotional representation (IP8)(M = 5.74, SD = 2.87), and also for mean scores of non-practitioner for personal control (M = 4.88, SD = 2.6), Coherence/understanding (IP7) (M = 6.83, SD = 2.28), emotional representation (IP8) (M = 5.66, SD = 2.97), all $ps<0.05$ with higher mean interpreted as a higher perception of personal control over the illness /COVID19, higher coherence/understanding and higher emotional representation of COVID19. However, there was no significant difference between the other spiritual practices and the non-practitioner group, all $ps >0.05$.

There was also a statistically significant difference in the mean scores of yoga practitioner for risk perception (IP10 and IP11) (M = 5.88, SD = 2.75), when compared with the mean scores of non-practitioner risk perception (M = 6.59, SD = 2.58), $p<0.05$, with lower mean scores interpreted as a lower perception of risk to contract COVID19. However, there was no significant difference in the mean scores of other spiritual practitioner group when compared with the yoga practitioner group, and the non-practitioner group, both $ps>0.05$.

There was also a statistically significant difference in the mean scores for the perception of preventive control (IP12) of yoga practitioner (M = 7.10, SD = 2.44), when compared with other spiritual practitioner group (M = 6.75, SD = 2.23), and non-practitioner (M = 6.17, SD = 2.62), $p<0.001$, with higher mean scores interpreted as a higher perception of personal preventive control over COVID19. However, there was no significant difference in the mean scores of the other spiritual practitioner group, when compared with the yoga practitioner group and the non-practitioner group, both $ps>0.05$. Means plot is shown in Fig 1.

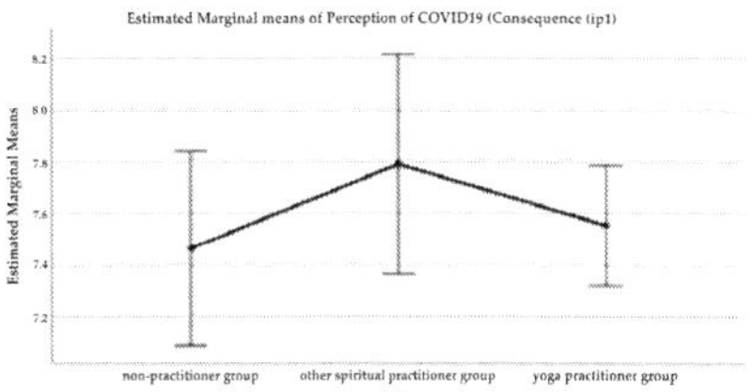

(i)

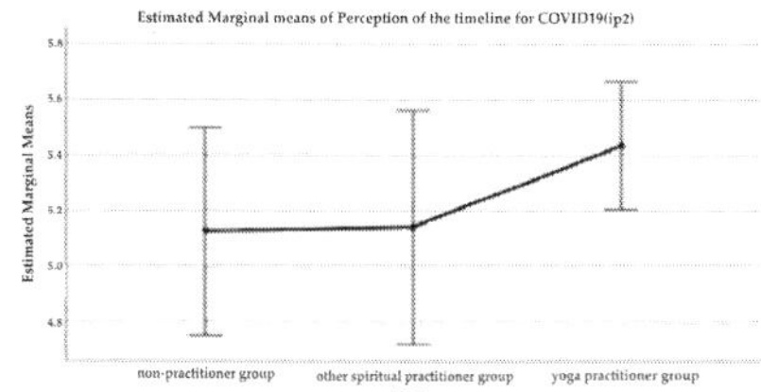

(ii)

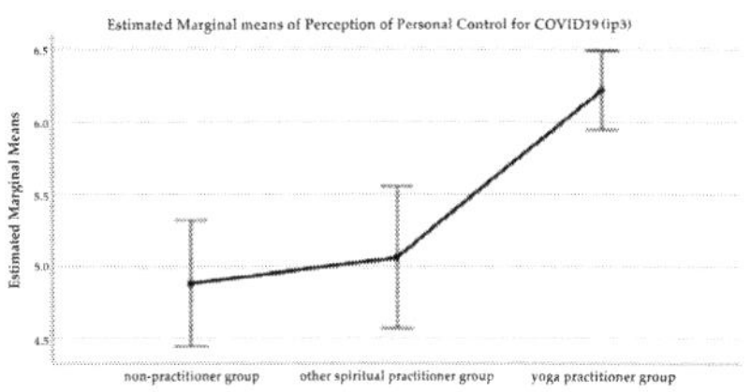

(iii)

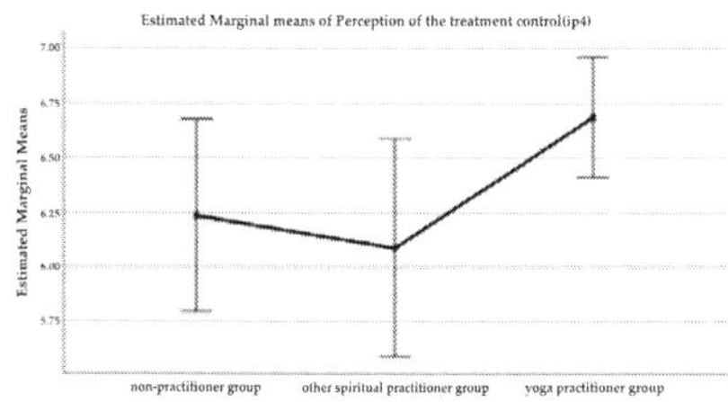

(iv)

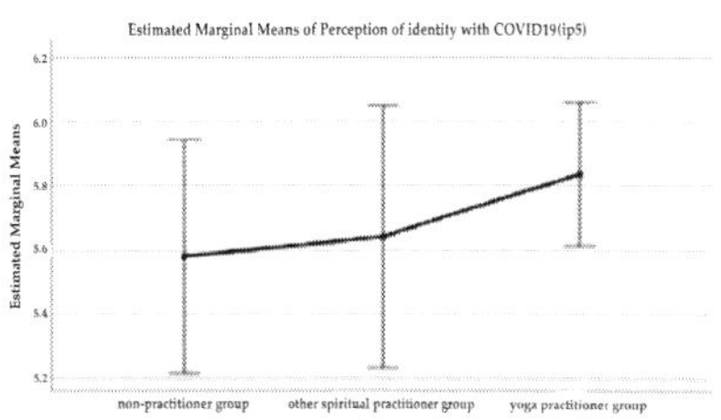

(v)

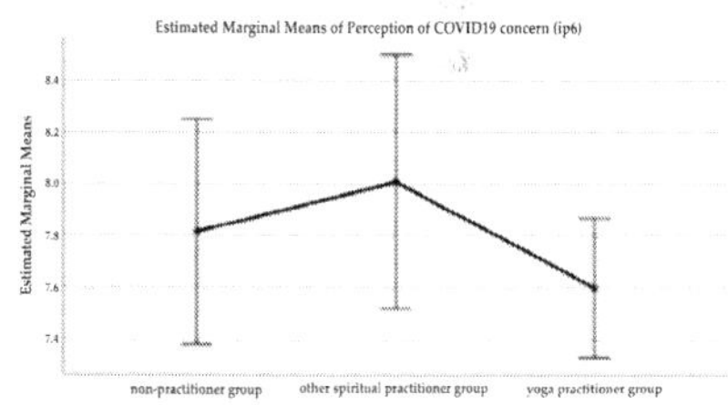

(vi)

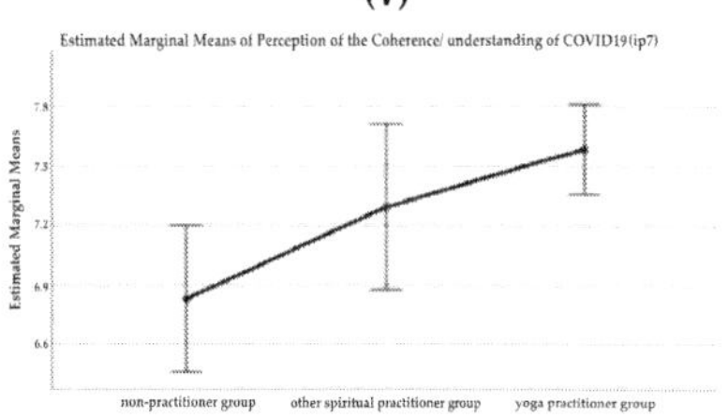

(vii)

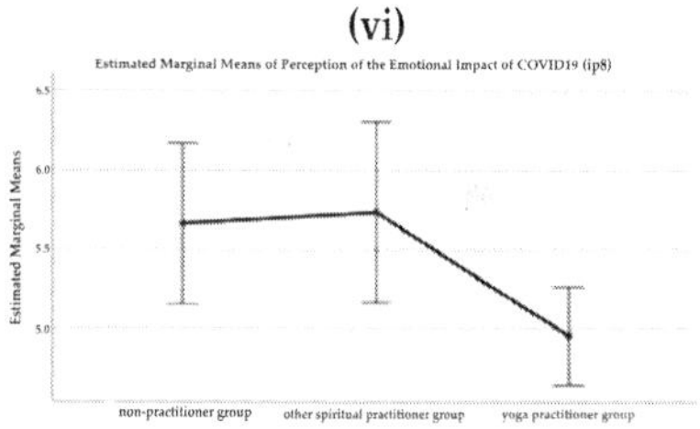

(viii)

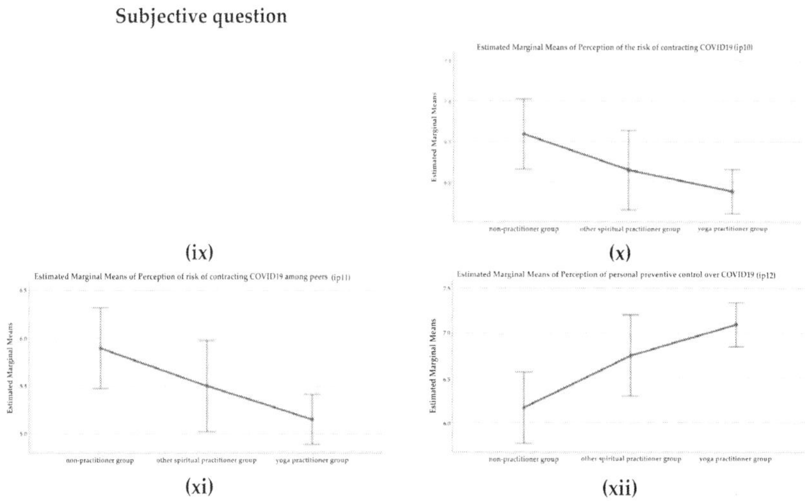

Figure 1. Means Plot for COVID19 perception of (i) consequence; (ii) timeline; (iii) personal control; (iv) treatment control; (v) identity; (vi) illness concern; (vii) coherence/ understanding; (viii) emotional representation; (x) risk perception; (xi) peer risk perception; (xii) preventive control for three practitioner groups. Note: Error bars at 95% Cl.

Percentage analysis of the causal representation of COVID19 on the responses received to the open-ended question (IP9) which asks patients to list the three most important causal factors in the rank order revealed that 48.25% ranked lack of adequate and timely information as the foremost reason for the spread of COVID19, followed by consumption of animal flesh (17.06%) and international movement of tourists and immigrants (16.74%). Other causal factors listed by participants included; lack of immunity, lack of medical facilities/treatment, improper/late action on lockdown measures etc.

Wellbeing Related Measure (DASS, SGWB, POMS, BRS)

There was a statistically significant difference in group means for depression (DASS-D) (F (2,639) = 12.48, $p < .001$, partial eta square = .058, and stress (DASS-S) (F (2,639) = 3.80, $p < .05$, partial eta square = .015. However, there was no statistically significant difference in group means for anxiety (DASS-A), p>.05.

Further, a post hoc analysis (Tukey HSD) revealed significant differences in mean scores for depression (DASS-D) among yoga practitioner group (M = 0.74, SD = .63), when compared with other spiritual practitioner group (M = .93, SD = .58), and non-practitioner group (M = 1.13, SD = .73), both ps< 0.001, with higher mean scores signifying higher depression. There was also a significant difference in mean scores for stress (DASS-S) for yoga practitioner group (M = .79, SD = .61) when compared with other spiritual practitioner group (M = .94, SD = .60), and non-practitioner group (M = .95, SD = .61), both ps< 0.001, with higher mean scores signifying higher stress. There was also a significant difference in the mean scores for stress (DASS-S) between other spiritual practitioners and the non- practitioner, p<0.05. Whereas, there was no statistically significant difference for anxiety or depression between the other spiritual practitioner group and non-practitioner group p>0.05.

Percentage analysis of the additional item on the three most important causal factors (in rank order) of stress during lockdown revealed that majority of the participants reported isolation due to lockdown (23.17%), fear of loss of job/business (financial insecurity) (18.57%), and fear of contracting virus (10.32%) as three main causes of stress during COVID19 lockdown. Other reasons of stress included; media reports and inadequate information (7.93%), the uncertainty of future (7.40%), routine disturbances (6.66%), educational loss (5.87%), and family issues (3.65%). Interestingly, 8.89% of the participants reported having no stress at all. A gender wise analysis of the causal factors of stress shows that females higher percentage of females (25.89%) reported stress due to isolation due to lockdown than males (22.11%). Whereas, financial insecurity was ranked as major cause of stress by more males (19.59%) than females (17.41%).

There was also a statistically significant difference in group means for wellbeing (SGWB) F (2,639) = 31.20, $p < .001$, partial eta squared = .112, and peace of mind (POMS) F (2,639) = 30.99, $p < .001$, partial eta square = .114. However, there was no statistically significant difference in group means for resilience (BRS), $p < .05$. Further, a post hoc analysis (Tukey HSD) revealed that mean scores of yoga practitioner group for wellbeing (SGWB) (M = 3.74, SD = .78), and peace of mind (POMS) (M = 3.565, SD = .96) differed significantly when compared with the other spiritual practitioner group mean scores for wellbeing (SGWB) (M = 3.28, SD = .67), peace of mind (POMS) (M = 3.13, SD = .96), and non-practitioner group mean scores for wellbeing (SWGB) (M = 3.11, SD = .85), peace of mind (POMS) (M = 2.73, SD = .99), all ps< 0.001. There was no significant difference between the mean scores for wellbeing (SWGB) between other spiritual practitioner groups and non-practitioner groups p>0.05. However, for peace of mind (POMS) there was a statistically significant difference between the other spiritual practitioner group and non-practitioner group p<0.001, with higher mean scores signifying higher wellbeing and higher peace of mind. Means plot is shown in Fig 2.

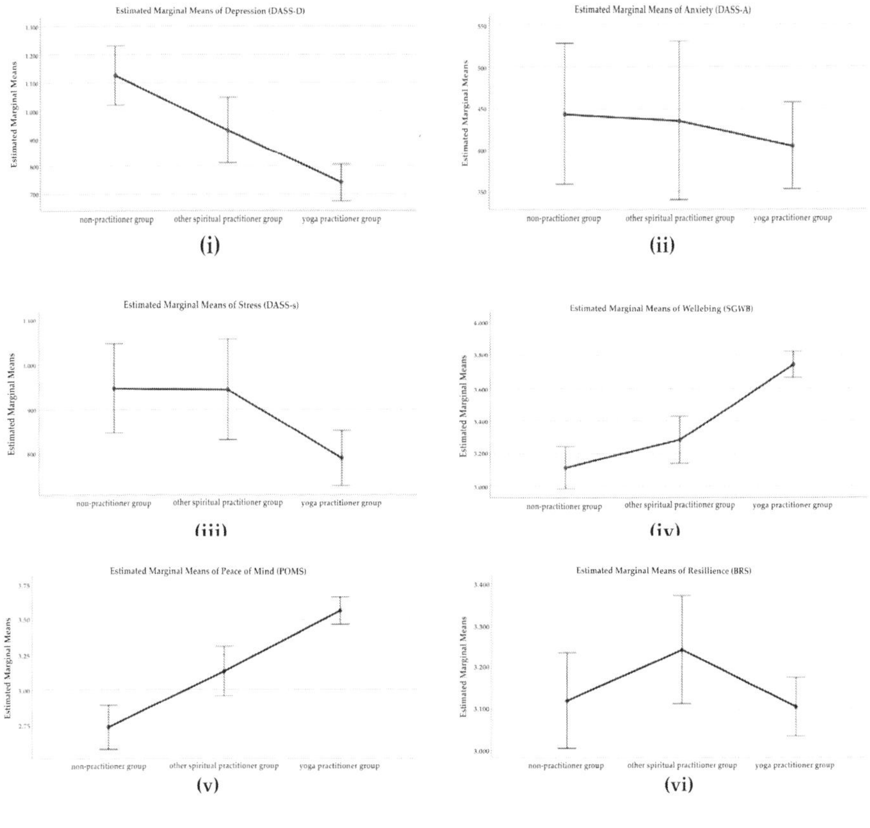

Figure 2. Means plot for Wellbeing measures (i) Depression; (ii) Anxiety; (iii) Stress; (iv) Wellbeing; (v) Peace of Mind and (vi) Resilience for three practitioner groups. Note: Error bars at 95% Cl.

Emotion Regulation Strategies (ERQ)

There was a statistically significant difference in group means for cognitive reappraisal strategies (ERQ-C) $F(2,639) = 14.85, p < .001$, partial eta square = .059. However, there was no significant difference in the group mean scores for Expressive suppression (ERQ-E), p>0.05). Further, a post hoc analysis (Tukey HSD) revealed significant differences in mean scores for yoga practitioner group (M = 5.24, SD = 1.07), other spiritual practitioner group (M = 4.85, SD = 1.17), and non-practitioner group (M = 4.57, SD = 1.19), both ps< 0.001. There was also a statistically significant difference between the other spiritual practitioner group and non-practitioner group p<0.001, with higher mean scores signifying higher emotion regulation through cognitive reappraisal strategies. Means plot is shown in Fig 3.

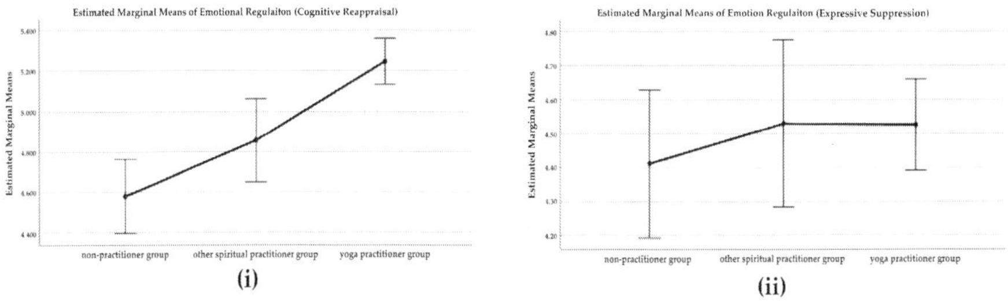

Figure 3. Means Plot for Emotion regulation i) Cognitive appraisal ii) Expressive suppression for three practitioner groups. Note: Error bars at 95% Cl.

Effect of Duration of Yoga Practice on Illness Perception and Wellbeing Measures

The MANOVA was conducted to test the hypothesis that there would be one or more mean differences in the wellbeing for different groups of yoga practitioners differentiated based on the number of practice years.

After controlling for the confounding effect of demographic variables a statistically significant MANOVA effect was obtained, Pillai's' Trace = .229, F (38, 716) = 2.43, p<0.001. The multivariate effect size was estimated at .114.

Before conducting a series of follow up ANOVAs, the homogeneity of variance assumption was tested for all the dependent variables. A series of Levene's F tests, to examine the homogeneity of variance assumption was conducted and is presented in Table 4.

A series of one-way ANOVA was conducted followed by a series of post hoc analyses (Tukey's HSD) were performed to examine the individual mean difference comparisons across all three groups of spiritual practitioners and all the dependent variables. The results revealed statistically significant comparisons as listed in Table 5.

Table 4. One-way ANOVA with wellbeing measures as the dependent variable and yoga practitioner groups based on the duration of practice as the independent variable.

	Levene's statistics		ANOVA		
	F	**Sig.**	**F**	**Sig.**	**Partial eta squared**
Consequence(IP1)	9.185	0.000	2.239	0.108	0.012
Timeline (IP2)	2.511	0.083	0.325	0.723	0.002
Personal control (IP3)	0.101	0.904	3.571	0.029	0.019
Treatment Control (IP4)	0.036	0.965	0.675	0.510	0.004
Identity (IP5)	0.172	0.842	0.292	0.747	0.002
Illness concern(IP6)	16.833	0.000	6.700	0.001	0.035
Coherence/understanding (IP7)	0.377	0.686	1.279	0.280	0.007
Emotional representation (IP8)	3.877	0.022	3.311	0.038	0.018
Risk perception (IP10)	4.312	0.014	1.535	0.217	0.008
Risk perception (IP11)	2.614	0.075	5.640	0.004	0.030
Personal preventative control(IP12)	0.359	0.699	1.428	0.241	0.008
Depression (DASS-D)	4.667	0.010	4.151	0.017	0.022
Anxiety (DASS-A)	7.746	0.001	5.189	0.006	0.027
Stress(DASS-S)	0.934	0.394	2.708	0.068	0.015
Peace of Mind (POMS)	1.232	0.293	15.100	0.000	0.076
Wellbeing (SGWB)	9.611	0.000	22.606	0.000	0.110
Resilience(BRS)	4.646	0.010	1.920	0.148	0.010
Emotion regulation-Cognitive Appraisal (ERQ-C)	0.834	0.435	5.298	0.005	0.028
Emotion regulation-(Expressive Suppression (ERQ-E)	1.373	0.255	2.337	0.098	0.013

Table 5. Summary of the post hoc analysis for the groups based on duration of yoga practice- long-term, mid- term, and beginner groups.

Measures	Long-term vs mid-term group	Mid-term vs beginner group	Long-term vs beginner group
Consequence(IP1)	n.s.	n.s.	n.s.
Timeline (IP2)	n.s.	n.s.	n.s.
Personal control (IP3)	n.s	p < .005	p < .005
Treatment Control (IP4)	n.s.	n.s.	n.s.
Identity (IP5)	n.s.	n.s.	n.s.
illness concern(IP6)	n.s.	p < .005	p < .005
coherence/understanding (IP7)	n.s.	n.s.	n.s.
Emotional representation (IP8)	n.s.	p < .005	p < .005
Risk perception (IP10)	n.s.	p < .005	p < .005
Peer Risk perception (IP11)	n.s.	p < .005	p < .005
Personal preventative control(IP12)	n.s.	p < .005	p < .005
Depression (DASS-D)	p<0.001	n.s.	p<0.001
Anxiety (DASS-A)	p<0.05	n.s.	p<0.05
Stress (DASS-S)	n.s.	n.s.	n.s.
Peace of Mind (POMS)	p<0.001	p<0.001	p<0.001
Wellbeing (SGWB)	p<0.05	n.s.	p<0.05
Resilience(BRS)	n.s.	n.s.	n.s.
Emotional regulation-Cognitive Appraisal (ERQ-C)		n.s.	p<0.05
Emotional regulation-(Expressive Suppression (ERQ-E)	n.s.	n.s.	n.s.

COVID19 Perception (BIPQ)

There was a statistically significant difference in group means for the Personal control (IP3) $F(2,367) = 3.571, p < .05$, partial eta squared = 0.02, Illness concern(IP6) $F(2,367) = 6.70, p < .001$, Emotional representation (IP8) $F(2,367) = 3.31, p < .0.05$, Risk perception(IP11) $F(2,367) = 5.64, p < .005$. However, there was no significant difference found in the mean scores of COVID19 representation of illness perception for the consequence (IP1), timeline (IP2), treatment control (IP4), identity (IP5) and Coherence/ Understanding (IP7)

A post hoc analysis (Tukey's HSD) revealed significant differences in mean scores of long term practitioner for personal control (IP3) (M = 6.68, SD = 2.79), illness concern(IP6) (M = 6.93, SD = 3.3), emotional representation (IP8) (M = 4.27, SD = 2.80), risk perception (IP10 and IP11) (M = 5.40, SD = 3.04) when compared with beginners mean scores of personal control (IP3) (M = 5.75, SD = 2.68),illness concern (M = 8.23, SD = 2.11), emotional representation (M = 5.44, SD = 2.89), risk perception(IP10 and IP11) ((M = 6.14, SD = 2.40), all ps<0.005, with higher mean interpreted as a higher perception of personal control, illness concern, emotional impact and higher risk perception of contracting COVID19. There was also a statistically significant difference in mean scores between beginners' illness concern (M = 8.23, SD = 2.11), and mid-term (M = 7.53, SD = 2.67) p<0.005. However, there was no significant difference between the long term and mid-term practitioners group for illness concern p>0.05. Means Plot shown in Fig 4.

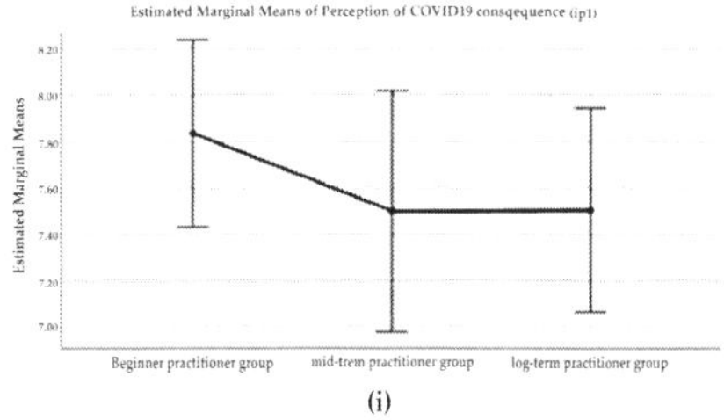

(i)

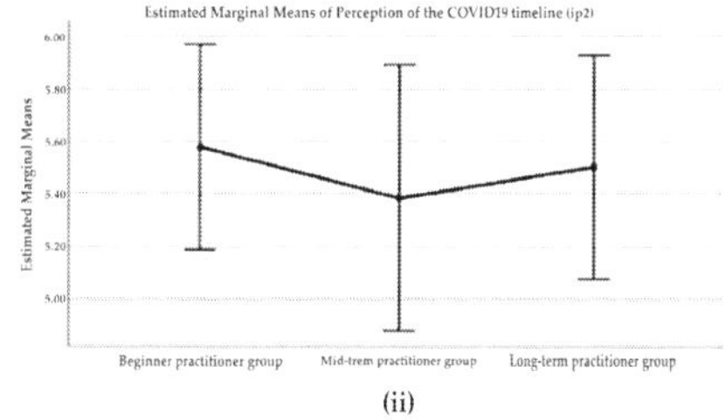

(ii)

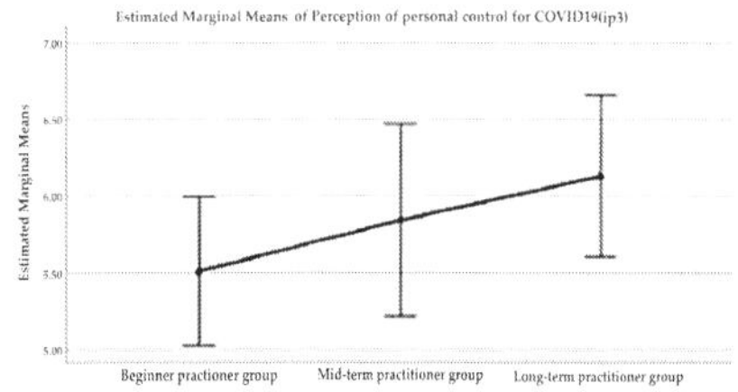

(iii)

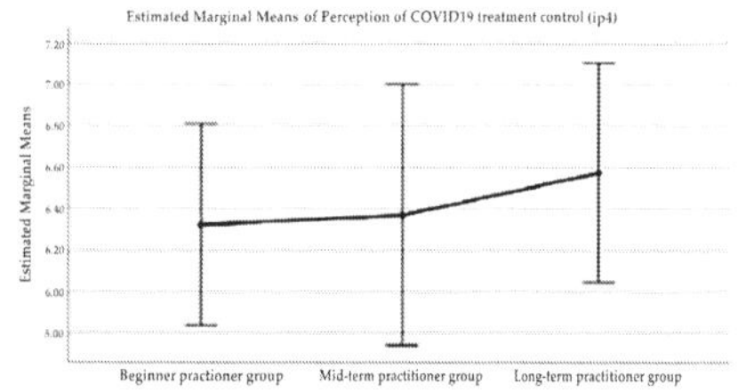

(iv)

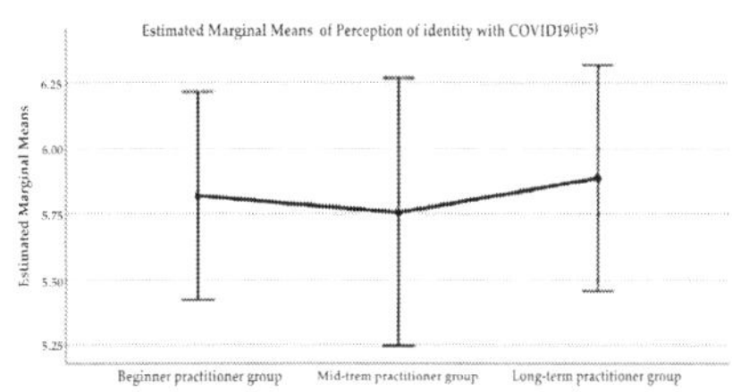

(v)

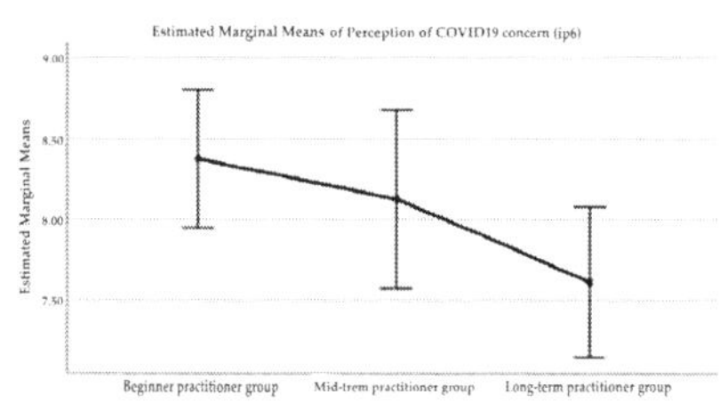

(vi)

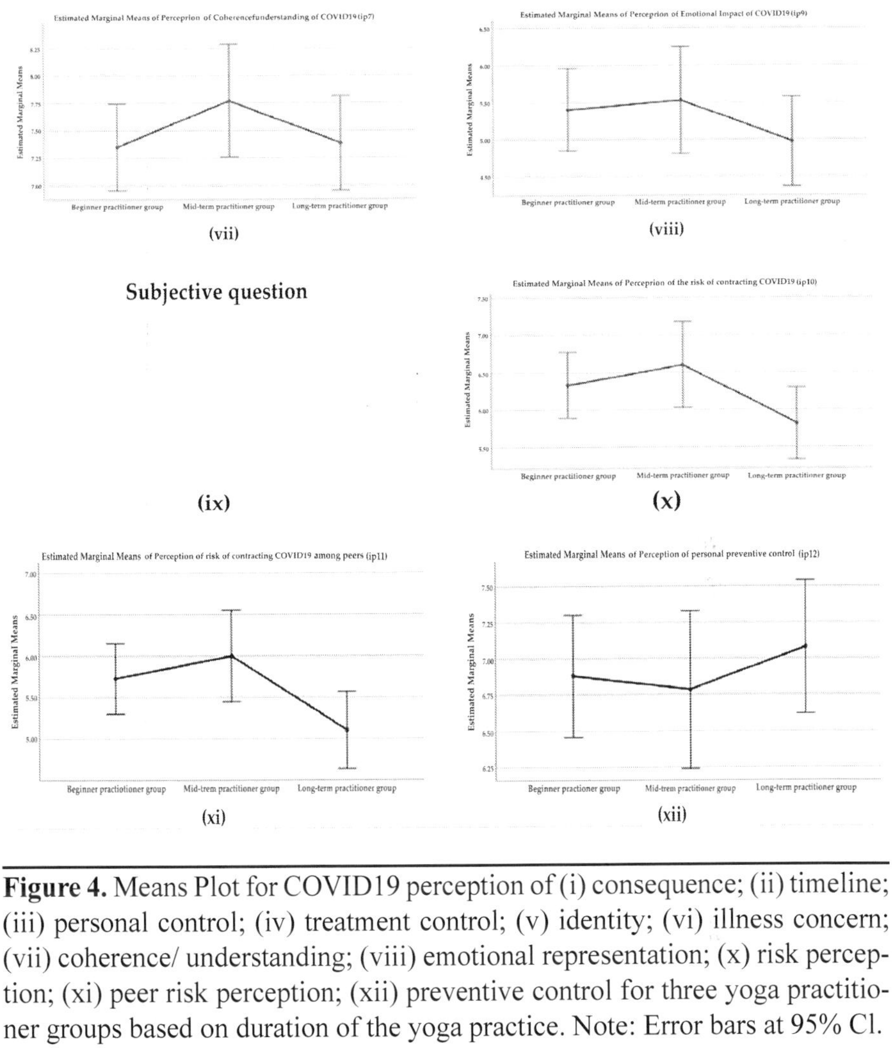

Figure 4. Means Plot for COVID19 perception of (i) consequence; (ii) timeline; (iii) personal control; (iv) treatment control; (v) identity; (vi) illness concern; (vii) coherence/ understanding; (viii) emotional representation; (x) risk perception; (xi) peer risk perception; (xii) preventive control for three yoga practitioner groups based on duration of the yoga practice. Note: Error bars at 95% Cl.

Wellbeing Related Measures (DASS, SGWB, POMS, BRS)

There was a statistically significant difference in group means for depression (DASS-D) (F (2,367) = 4.15, $p < .05$, partial eta square = .022, anxiety (DASS-A) F (2,367) = 5.19, $p < .01$, partial eta square = .03 and a trend in group means for stress (DASS-S), F (2,367) = 2.71, $p = .068$. Further, a post hoc analysis (Tukey HSD) revealed significant differences in mean scores between long term practitioner group depression (DASS-D)(M = 0.55, SD = .55), Anxiety (DASS)(M = .26, SD = .38), and mid-term practitioner group depression (DASS-D) (M = .81, SD = .67), anxiety (DASS-A) (M = .48,

SD = .56), and beginners group depression (DASS-D) (M = .80, SD = .60), all ps< 0.005, with higher mean scores signifying higher depression, and anxiety. There was no statistically significant difference between the mid-term and beginner practitioner group.

There was a statistically significant difference in group means for SGWB (F (2,367) = 22.60, p < .001, partial eta square = .110, and POMS (F (2,375) = 15.10, p < .001, partial eta square = .076. However, there was no statistically significant difference found in group means for BRS, p < .05. Post hoc analysis (Tukey HSD) revealed significant differences in mean scores of Wellbeing (SGWB) of long term practitioner group (M = 4.10, SD = .64) when compared with mid- term practitioner group (M = 3.79, SD = .65) and beginners group (M = 3.42, SD = .84), both ps< 0.001. Mean scores of POMS for long-term practitioner group (M = 4.03, SD = .91) also differed significantly with the mean scores of mid-term practitioner group (M = 3.50, SD = .87), and beginner practitioner group (M = 3.27, SD = .94), both ps< 0.001. There was also a statistically significant difference in the wellbeing (SGWB) mean scores and peace of mind (POMS) mean scores between the mid-term practitioner group and beginner group p>0.05, with higher mean scores signifying higher peace of mind. Means plot is shown in Fig 5.

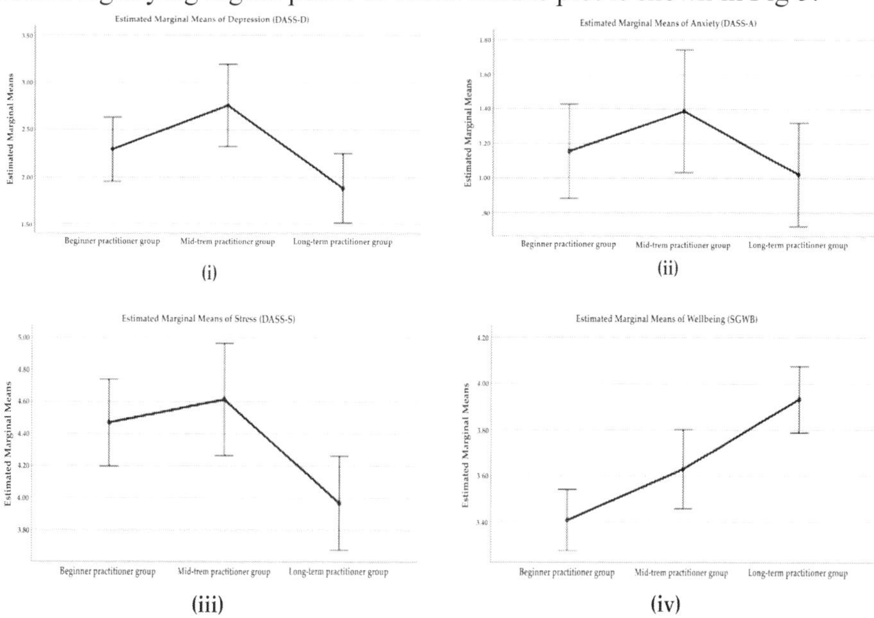

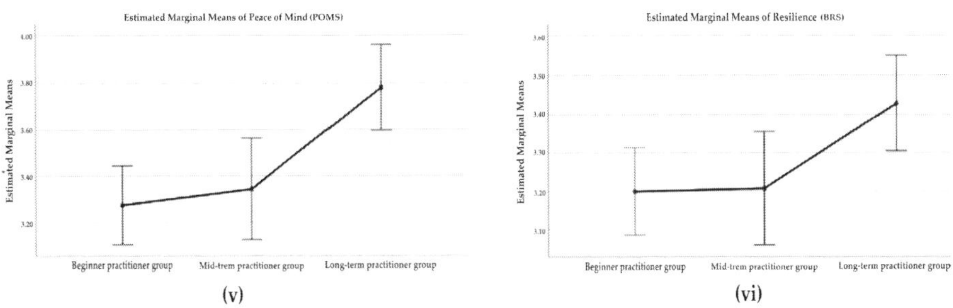

Figure 5. Means plot for Wellbeing measures (i) Depression; (ii) Anxiety; (iii) Stress; (iv) Wellbeing; (v) Peace of Mind and (vi) Resilience for three yoga practitioner groups based on duration of the yoga practice. Note: Error bars at 95% Cl.

Emotion Regulation Measure (ERQ)

There was a statistically significant difference in group means for ERQ (cognitive reappraisal) (F (2,375) = 5.30, p < .005, partial eta square = .028. However, there was no significant difference in the group mean scores for ERQ (Expressive suppression), p>0.05), indicating that the duration of yoga practice affects cognitive reappraisal strategies for emotion regulation. Further, a post hoc analysis (Tukey HSD) revealed significant differences in mean scores between long-term practitioner group (M = 5.51, SD = .97), and beginner group (M = 5.0, SD = 1.02) with higher mean scores signifying higher emotion regulation through cognitive reappraisal strategies. There was no statistically significant difference between the mid-term practitioner group and beginner group p>0.05. Means plot is shown in Fig 6.

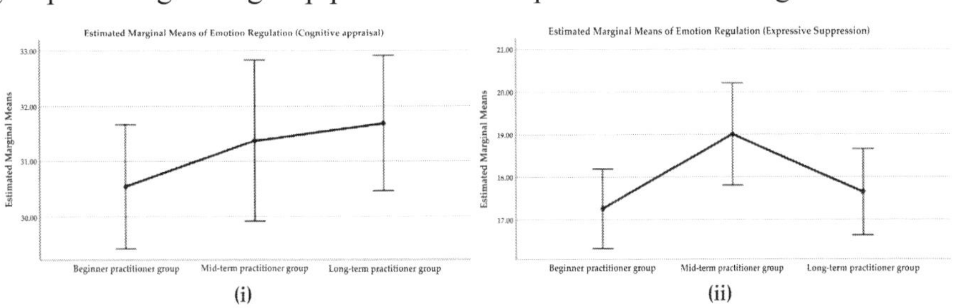

Figure 6. Means Plot for Emotion regulation i) Cognitive appraisal ii) Expressive suppression for three practitioner groups. Note: Error bars at 95% Cl.

DISCUSSION AND CONCLUSIONS

The aim of present research was to study the effect of practice of yoga and other spiritual practices on the illness perception, wellbeing measures and emotion regulation strategies for adults during COVID19 lockdown. Additionally, the effect of demographic variables such as age, gender, qualification, working status and place of residence was also analyzed. The confirmatory factor analysis confirmed the factor structure for the study sample, which strengthens the findings of this research.

The results examining the demographic variables demonstrate that females reported higher illness concern and were emotionally more impacted than males by the COVID19 lockdown. This finding is in line with an earlier review of literature that reported females to be twice more vulnerable to stress than males in conditions or events of adversity [16, 17, 19]. A recent study investigating depression, anxiety and stress due to COVID19 also found higher stress among females than males [42]. The reason for higher stress in females can be argued to be partially due to increase in the household chores in the absence of any house helps during lockdown, specially for those also managing their professional work. Further, it was also found that younger participants from age group 18–25 years reported feeling more depressed, and had lower peace of mind than older participants.

A stressful situation such as fear of losing a job, uncertainty about the future can trigger anxiety, depression and is believed to affect the peace of mind. The rationale for the association of stress and age is given by a study investigating differences in coping strategies across lifespan. The study suggests that older adults use coping strategies that are indicative of greater impulse control and they tend to evaluate conflict situations more positively than younger adults [16–18]. In another study it was found that older adults had lower levels of psychological distress and better dispositional coping compared to younger adults [43]. Perhaps the fear arising from uncertainties was dealt more efficiently by the older population, such that it affected their wellbeing positively. Interestingly, in this study the urban population reported higher perception of the COVID19 consequences than reported by the participants from rural or semi-urban areas. Perhaps, the urban population felt that COVID19 lockdown is going to affect them more adversely than the rural or semi-urban. One of the reasons for this difference could be the job insecurity. Another plausible explanation for the difference in perception of the COVID 19 consequence can be derived from research that states that urban populations are reported more prone to psychological distress than their rural counterparts [44]. Since urban population consists of the majority of service class which is dependent on their jobs for livelihood, they are more likely to perceive graver consequences of COVID19 lockdown than their rural counterparts which comprise mostly of self- employed people.

The results examining the effect of yoga practice demonstrate that yoga practitioners perceived having higher personal control, higher coherence/understanding, lower emotional impact, lower risk and higher preventive control for contracting COVID19 than other spiritual practitioners and non-practitioners. A number of studies have reported physical and mental health benefits of yoga practice [45–47].

A healthy individual is found to perceive lower cognitive and emotional effects of the illness and a higher preventive control over the illness [9]. On the other hand, the participants who negatively perceived the COVID19 effects experienced greater levels of stress, anxiety or depression and lower wellbeing, also reported in a study on cancer patients [48]. Additionally, in light of the findings of the previous study, the notion that an individual is following a treatment or preventative control therapy positively affects the perception about how well the illness is understood and a sense of personal control over the illness [10]. In this study also yoga practitioners reported to have a better understanding and higher personal control over COVID19. Perhaps yoga practitioners felt that yoga is an effective therapy to cope with COVID19 both for physical as well as mental health.

In this study, it was also found that yoga practitioners had lower depression, lower stress, lower anxiety, higher wellbeing, and higher peace of mind than the other spiritual practitioners and non-practitioner group. Interestingly, the other spiritual practitioners were also found to have a significantly higher peace of mind than the non-practitioners. The other spiritual practitioner group also reported lower depression, anxiety, stress and higher wellbeing than the non-practitioner group, however the difference was not found to be statistically significant in this study. Possibly the other spiritual practices; reading Holy Scriptures and rendering *seva* (selfless service) to the needy and destitute provided solace and peace of mind. A previous study has also reported a positive association between reading scriptures and positive affect and *sukha* (happiness) and a negative association with negative affect and *dukkha* (unhappiness) [32]. As for the non-practitioner group, participants that reportedly followed none of the yoga or spiritual practices, also reported the highest mean score of depression, anxiety and stress and lowest wellbeing and peace of mind.

Results showed that there was no significant difference in resilience among the yoga practitioners, other spiritual practitioners and non-practitioner group. Resilience has for long been debated by researchers to be a trait construct. In this study also resilience was found to be perhaps a more trait-like construct that unfolds over time in response to internal strengths and external supports across lifespan that foster positive outcomes in the face of adversity.

In this study, a significant effect of duration of practice was found on illness perception, and wellbeing related measures. Long term practitioners reported higher personal control and lower illness concern in contracting COVID19 than the mid-term or beginner group. The long- term and mid-term practitioners also reported perceiving lower emotional impact of COVID19 and lower risk in contracting COVID19 than the beginners. The general wellbeing was reported higher by the long term and mid- term practitioners than the beginners group. Further, the long term practitioners were found to have highest peace of mind, lowest depression and anxiety with no significant difference in the mid-term and the beginner group. In an earlier study, sustained practice of yoga is reported to enhance physical strength, promote and improve respiratory and cardiovascular function. The improved physiological functions are believed to reduce stress, anxiety, depression, and enhance overall well-being. In line with the outcomes from this study, the regular practice has also been argued to lead changes in life perspective, self-awareness, a sense of balance between body and mind and generally a positive outlook to life that maintains general wellbeing even in difficulties [49, 50]. Interestingly, in this study the beginner group, which had started practicing yoga during COVID19 lockdown, reported comparable mean scores of wellbeing and peace of mind with the mid-term practitioner groups. When compared with the non-practitioner group, the beginner group also had lower depression, anxiety, stress and higher wellbeing, peace of mind. Perhaps the routine practice of yoga helped the beginner practitioners to calm the mind and maintain a positive disposition during difficult times of COVID19 lockdown. The outcomes reveal that yoga practice helps in illness perception about COVID19 such that the long- term practitioners feel a better sense of preventive control with a notion of being less prone to contracting COVID19. This perception of lesser vulnerability and a better sense of control over COVID19 is argued to generate lesser stress problems and promote higher wellbeing. The emotion regulation strategy of cognitive reappraisal is further argued to breed a balanced and coherent understanding about the COVID19. The balanced representation of the unknown is argued to tone down the fear factor due to uncertainties caused by COVID19 lockdown thus decreasing the stress, anxiety and depression. Such a state of mind allows one to view an adverse situation with a more pragmatic approach and helps in maintaining a peaceful disposition.

Altogether, the findings from this study shows that yoga is found to be an effective way to manage the stress, anxiety and depression due to and during COVID19 lockdown. The evidence further supports that yoga could be used as a complementary and alternative therapy for the stress related problems due to COVID19. It may also help health practitioners in further promoting yoga-based interventions to facilitate the self-management of the mental health issues due to COVID19.

SUPPORTING INFORMATION

S1 Table. Correlations

		1	2	3	4	5	6	7	8	9	10	11	12	13	14	15	16	17	18	19	Mean	Std. Deviation
1	Consequence (IP1)	1																			7.58	2.32
2	Timeline (IP2)	.283**	1																		5.31	2.29
3	Personal control (IP3)	-0.012	0.064	1																	5.71	2.75
4	Treatment Control (IP4)	.107**	0.048	.315**	1																6.48	2.71
5	Identity (IP5)	.191**	.191**	.172**	.175**	1															5.75	2.23
6	Illness concern(IP6)	.385**	.225**	-0.048	.142**	.220**	1														7.72	2.66
7	Coherence/ understanding (IP7)	0.074	0.077	.268**	.158**	.201**	.261**	1													7.36	2.29
8	Emotional representation (IP8)	.163**	.171**	-.143**	.162**	.133**	.407**	.117**	1												5.26	3.11
9	Risk perception (IP10)	.154**	.122**	-.172**	-0.002	.216**	.267**	0.057	.234**	1											6.09	2.68
10	Risk perception (IP11)	.136**	.156**	-.193**	-0.016	.165**	.293**	0.037	.221**	.734**	1										5.38	2.61
11	Personal preventative control(IP12)	0.031	.083*	.361**	.221**	.156**	0.067	.319**	-0.015	-.124**	-.169**	1									6.83	2.47

12	Depression (DASS-D)	0.052	0.028	-.135**	-0.017	0.033	0.005	-.086*	.197**	.148**	.130**	-.151**	1								0.86	0.66
13	Anxiety (DASS-A)	0.010	.117**	-0.041	0.050	.104**	0.043	-0.006	.301**	.156**	.168**	-0.076	.430**	1							0.42	0.52
14	Stress (DASS-S)	.079*	0.036	-.116**	-0.011	.090*	0.063	-0.071	.182**	.091*	.111**	-.088*	.540**	.533**	1						0.85	0.62
15	Peace of Mind (POMS)	0.006	0.022	.303**	.194**	0.071	-0.011	.278**	-.203**	-.118**	-.138**	.269**	-.473**	-.371**	-.374**	1					3.30	1.03
16	Wellbeing (SGWB)	0.039	0.016	.318**	.191**	.112**	0.029	.288**	-.156**	-.101*	-.130**	.312**	-.460**	-.278**	-.259**	.734**	1				3.52	0.83
17	Resilience (BRS)	0.022	0.022	0.044	-0.064	0.038	.134**	.080*	-0.058	0.020	0.070	.136**	-.186**	-.195**	-.102*	.173**	.199**	1			3.13	0.70
18	Emotion regulation-Cognitive Appraisal (ERQ-C)	0.068	0.057	.239**	.179**	.091*	.121**	.300**	0.006	-0.019	0.003	.308**	-.202**	-0.066	-.097*	.378**	.448**	.166**	1		5.02	1.15
19	Emotion regulation-(Expressive Suppression (ERQ-E)	0.051	.087*	.148**	.155**	0.073	.079*	.130**	.102**	0.038	0.028	.147**	.084*	0.054	-0.033	0.074	.085*	-0.052	.259**	1	4.50	1.34

ACKNOWLEDGMENTS

The authors appreciates all those who participated in the study and helped to facilitate the research process.

REFERENCES

1. Website. [cited 25 Sep 2020]. Available: Van Hoof, E. (2020, April). Lockdown is the world's biggest psychological experiment-and we will pay the price. In World Economic Forum (Vol. 9). https://www.weforum. org/agenda/2020/04/this-is-the-psychological-side-of-the-COVID19-pandemic-that-were-ignoring
2. Grover S, Sahoo S, Mehra A, Avasthi A, Tripathi A, Subramanyan A, et al. Psychological impact of COVID-19 lockdown: An online survey from India. Indian Journal of Psychiatry. 2020. p. 354. pmid:33165368
3. Boyraz G, Legros DN. Coronavirus Disease (COVID-19) and Traumatic Stress: Probable Risk Factors and Correlates of Posttraumatic Stress Disorder. Journal of Loss and Trauma. 2020. pp. 503–522.
4. Petrie KJ, Cameron LD, Ellis CJ, Buick D, Weinman J. Changing illness perceptions after myocardial infarction: an early intervention randomized controlled trial. Psychosom Med. 2002;64: 580–586. pmid:12140347
5. Leventhal H, Nerenz DR, Steele DJ. Illness Representations and Coping With Health Threats. Handbook of Psychology and Health (Volume IV). 2020. pp. 219–252.
6. Williams L, O'Connor RC, Grubb NR, O'Carroll RE. Type D personality and illness perceptions in myocardial infarction patients. J Psychosom Res. 2011;70: 141–144. pmid:21262416
7. Lau RR, Hartman KA. Common sense representations of common illnesses. Health Psychology. 1983. pp. 167–185.
8. Broadbent E, Petrie KJ, Main J, Weinman J. The brief illness perception questionnaire. J Psychosom Res. 2006;60: 631–637. pmid:16731240
9. Website. [cited 25 Sep 2020]. Available: Vos R. C. , Kasteleyn M. J. , Heijmans M. J. , de Leeuw E. , Schellevis F. G. , Rijken M. ,. . . & Diacourse study group. (2018). Disentangling the effect of illness perceptions on health status in people with type 2 diabetes after an acute coronary event. BMC family practice, 19(1), 35. pmid:29499658
10. Lamiani G, Strada I, Mancuso ME, Coppola A, Vegni E, Moja EA, et al. Factors influencing illness representations and perceived adherence in haemophilic patients: a pilot study. Haemophilia. 2015;21: 598–604. pmid:25684356
11. Grossi E. , Tavano Blessi G. & Sacco P. L. (2019) Magic Moments: Determinants of Stress Relief and Subjective Wellbeing from Visiting a Cultural Heritage Site. Cult Med Psychiatry 43, 4–24. pmid:29998384

12. Selamu M., Thornicroft G., Fekadu A. et al. (2017) Conceptualization of job-related wellbeing, stress and burnout among healthcare workers in rural Ethiopia: a qualitative study. BMC Health Serv Res 17, 412. pmid:28629360
13. Davis T. J., Morris M., Drake M. M. (2016). The moderation effect of mindfulness on the relationship between adult attachment and wellbeing, Personality and Individual Differences,(96),115–121, https://doi.org/10. 1016/j. paid. 2016. 02. 080.
14. Zammuner VL, Galli C. Wellbeing: causes and consequences of emotion regulation in work settings. Int Rev Psychiatry. 2005;17: 355–364. pmid:16194815
15. Lee Y-C, Lin Y-C, Huang C-L, Fredrickson BL. Peace of Mind Scale. PsycTESTS Dataset. 2014.
16. Grant AM, Curtayne L, Burton G. Executive coaching enhances goal attainment, resilience and workplace well-being: a randomised controlled study. The Journal of Positive Psychology. 2009. pp. 396–407.
17. Woodyard C. (2011). Exploring the therapeutic effects of yoga and its ability to increase quality of life. International journal of yoga, 4(2), 49–54. pmid:22022122
18. Bhatia T. (2020). International Yoga Day 2020: Why Practising Yoga is Significant during COVID19 Pandemic?https://www. india.com/lifestyle/international-yoga-day-2020-why-practicing-yoga-is-significant-during-COVID19-pandemic-4063858/
19. Nagendra H. R. (2020). Yoga for COVID-19. International journal of yoga, 13(2), 87–88. pmid:32669761
20. Büssing A, Michalsen A, Khalsa SBS, Telles S, Sherman KJ. Effects of Yoga on Mental and Physical Health: A Short Summary of Reviews. Evidence-Based Complementary and Alternative Medicine. 2012. pp. 1–7. pmid:23008738
21. Cowen VS, Adams TB. Physical and perceptual benefits of yoga asana practice: results of a pilot study. Journal of Bodywork and Movement Therapies. 2005. pp. 211–219.
22. Goyal R., Lata H., Walia L., &Narula M. K. (2014). Effect of pranayama on rate pressure product in mild hypertensives. International Journal of Applied and Basic Medical Research, 4(2), 67. pmid:25143878

23. http://paperpile. com/b/khVgDP/tYSXBhushan S. (2003), Pratyahara Practice as a Method of Managing Negative Mental Health. YOGA. http://www. yogamag. net/archives/2003/cmay03/prat. shtml
24. Singh K. , Jain A. , & Singh D. (2014) Satsang: A Culture Specific Effective Practice for Well-Being. In: ÁguedaMarujo H. , Neto L. (eds) Positive Nations and Communities. Cross-Cultural Advancements in Positive Psychology, vol 6. 79–100 Springer, Dordrecht. https://doi. org/10. 1007/978-94-007-6869-7_5
25. Singh K, Sahni P. Swadhyaya Scale: An Indian Perspective. The International Journal of Indian Psychology. 2016;4: 5–14. pmid:27994352
26. Moreira-Almeida A, Neto FL, Koenig HG. Religiousness and mental health: a review. Braz J Psychiatry. 2006;28: 242–250. pmid:16924349
27. Lepherd L, Rogers C, Egan R, Towler H, Graham C, Nagle A, et al. Exploring spirituality with older people: (1) rich experiences. J Relig Spiritual Aging. 2020;32: 306–340.
28. Yusoff MSB, Yaacob MJ, Naing NN, Esa AR. Psychometric properties of the Medical Student Well-Being Index among medical students in a Malaysian medical school. Asian J Psychiatr. 2013;6: 60–65. pmid:23380320
29. Lovibond PF, Lovibond SH. The structure of negative emotional states: Comparison of the Depression Anxiety Stress Scales (DASS) with the Beck Depression and Anxiety Inventories. Behaviour Research and Therapy. 1995. pp. 335–343. pmid:7726811
30. Website. [cited 25 Sep 2020]. Available: Longo Y. , Coyne I. , & Joseph S. (2017). The scales of general well-being (SGWB). Personality and Individual Differences, 109, 148–159. https://doi. org/10. 1016/j. paid. 2017. 01. 005
31. Smith BW, Dalen J, Wiggins K, Tooley E, Christopher P, Bernard J. The brief resilience scale: assessing the ability to bounce back. Int J Behav Med. 2008;15: 194–200. pmid:18696313
32. Singh K, Mitra S, Khanna P. Psychometric properties of hindi version of peace of mind, harmony in life and sat-chit-ananda scales. Indian J Clin Psychol. 2016;43: 58–64.
33. Gross JJ, John OP. Emotion Regulation Questionnaire. PsycTESTS Dataset. 2012.

34. Gross JJ, John OP. Individual differences in two emotion regulation processes: implications for affect, relationships, and well-being. J Pers Soc Psychol. 2003;85: 348–362. pmid:12916575
35. Iglewicz B, Hoaglin DC. Use of Boxplots for Process Evaluation. Journal of Quality Technology. 1987. pp. 180–190.
36. Geuens M, De Pelsmacker P. Developing a Short Affect Intensity Scale. Psychol Rep. 2002;91: 657–670. pmid:12416861
37. MacCallum RC, Browne MW, Sugawara HM. Power analysis and determination of sample size for covariance structure modeling. Psychol Methods. 1: 130–149.
38. Hu L, Bentler PM. Cutoff criteria for fit indexes in covariance structure analysis: Conventional criteria versus new alternatives. Struct Equ Modeling. 1999;6: 1–55.
39. Tabachnick BG, Fidell LS, Ullman JB. Using multivariate statistics. Pearson Boston, MA; 2007.
40. Meyers LS, Gamst GC, Guarino AJ. Performing Data Analysis Using IBM SPSS. John Wiley & Sons; 2013.
41. Huberty CJ, Petoskey MD. Multivariate Analysis of Variance and Covariance. Handbook of Applied Multivariate Statistics and Mathematical Modeling. 2000. pp. 183–208.
42. Rehman U, Shahnawaz MG, Khan NH, Kharshiing KD, Khursheed M, Gupta K, et al. Depression, Anxiety and Stress Among Indians in Times of Covid-19 Lockdown. Community Ment Health J. 2020. pmid:32577997
43. Segal DL, Coolidge FL, Mizuno H. Defense mechanism differences between younger and older adults: A cross-sectional investigation. Aging & Mental Health. 2007. pp. 415–422. pmid:17612805
44. Stickley A, Koyanagi A, Roberts B, McKee M. Urban–rural differences in psychological distress in nine countries of the former Soviet Union. Journal of Affective Disorders. 2015. pp. 142–148. pmid:25813456
45. Kirkwood G. Yoga for anxiety: a systematic review of the research evidence * Commentary. British Journal of Sports Medicine. 2005. pp. 884–891. pmid:16306493
46. Smith C, Hancock H, Blake-Mortimer J, Eckert K. A randomised comparative trial of yoga and relaxation to reduce stress and anxiety. Complementary Therapies in Medicine. 2007. pp. 77–83. pmid:17544857

47. Shaikh S, Kumar S. A Comparative Study between Relaxation Technique versus 12 Moves of Yoga on Anxiety in Young Adults—A Randomized Clinical Trial. Indian Journal of Physiotherapy and Occupational Therapy—An International Journal. 2013. p. 207.
48. Miceli J, Geller D, Tsung A, Hecht CL, Wang Y, Pathak R, et al. Illness perceptions and perceived stress in patients with advanced gastrointestinal cancer. Psychooncology. 2019;28: 1513–1519. pmid:31090125
49. Desikachar K, Bragdon L, Bossart C. The Yoga of Healing: Exploring Yoga's Holistic Model for Health and Well-being. International Journal of Yoga Therapy. 2005. pp. 17–39.
50. Benefits Atkinson N. , Barriers, and Cues to Action of Yoga Practice: A Focus Group Approach. American Journal of Health Behavior. 2009. pmid:18844516

Chapter 5

Effect of an 8-Week Yoga-Based Lifestyle Intervention on Psycho-Neuro-Immune Axis, Disease Activity, and Perceived Quality of Life in Rheumatoid Arthritis Patients: A Randomized Controlled Trial

Surabhi Gautam[1], Manoj Kumar[1], Uma Kumar[2] and Rima Dada[1]

[1] Lab for Molecular Reproduction and Genetics, Department of Anatomy, All India Institute of Medical Sciences, New Delhi, India,

[2] Department of Rheumatology, All India Institute of Medical Sciences, New Delhi, India

ABSTRACT

Various external stressors and environmental challenges lead to the provocation of the immune system in autoimmune diseases like Rheumatoid

Citation: Gautam S, Kumar M, Kumar U and Dada R (2020) Effect of an 8-Week Yoga-Based Lifestyle Intervention on Psycho-Neuro-Immune Axis, Disease Activity, and Perceived Quality of Life in Rheumatoid Arthritis Patients: A Randomized Controlled Trial. Front. Psychol. 11:2259. doi: 10.3389/fpsyg.2020.02259 arthritis (RA). The inappropriate immune response further triggers the cascade of inflammatory changes resulting in precipitation of symptoms and hampers quality of life (QOL). The underlying psycho-somatic component of the disease requires a holistic approach to its treatment dimension rather than the use of pharmacotherapy. The applicability of mind-body interventions has become essential in today's fast-paced life. Yoga, a mind-body technique, alters the mind's capacity to facilitate systemic functioning at multiple organ system levels. Hence, we conducted this study to evaluate the impact of 8 weeks of a yoga-based lifestyle intervention (YBLI) on psycho-neuro-immune markers, gene expression patterns, and QOL in RA patients on routine medical therapy. A total of 66 patients were randomized into two groups: yoga group or non-yoga group and were assessed for a panel of inflammatory cytokines (IL-6, IL-17A, TNF-α, and TGF-β), mind-body communicative markers (BDNF, DHEAS, β-endorphin, and sirtuin) and transcript levels of various genes (*IL-6, TNF*-α, *NFKB1, TGF*-β, and *CTLA4*). We assessed disease activity and QOL using the DAS28-ESR and WHOQOL-BREF questionnaire, respectively. Yoga group observed significant improvements in the levels of markers, which influenced the psycho-neuro-immune axis ($p < 0.001$) with an estimated effect size from small to medium range. In the yoga group, there was a significant reduction in DAS28-ESR ($p < 0.001$) and improvement seen in the physical health, psychological, social relationships domains ($p < 0.001$) of QOL, except environmental ($p > 0.05$). The yoga group showed downregulation of *IL-6, TNF*-α, and *CTLA4* and upregulation of *TGF*-β. These results suggest that a decrease in disease activity after yoga practice is associated with a significant reduction in inflammatory cytokines, the elevation of mind-body communicative markers, and normalization of various transcript levels, which improved QOL. Thus the adoption of YBLI improves clinical outcome in RA, and decreases systemic inflammation by its beneficial effects on psycho-neuro-immune axis and normalization of dysregulated transcripts. Thus YBLI may be used for RA patients as an adjunctive therapy.

Keywords :- Rheumatoid arthritis, Quality of life, Inflammation, Disease activity score, Yoga-based lifestyle intervention, Psycho-neuro-immune axis

INTRODUCTION

Rheumatoid arthritis (RA) is a progressive autoimmune inflammatory disease characterized by arthritis of synovial-lined joints with variable systemic manifestation (Guo et al., 2018). It is the most commonly diagnosed systemic inflammatory arthritis, which severely impacts the quality of life (QOL) (Malm et al., 2017). Previous studies have shown an association of psychological stress with the onset and propagation of RA spanning both physical and mental domains of well-being (Margaretten et al., 2011; Matcham et al., 2013).

Various articular and systemic manifestations in RA are associated with disability and reduced life expectancy (Cimmino et al., 2000). Multiple factors like loss of work productivity increased work disability, regular health care visits, long term treatment with disease-modifying anti-rheumatic drugs (DMARDs), and the need for joint replacement surgeries are associated with RA. They pose a significant burden on the individual, caregivers, and health care system (van Vilsteren et al., 2015).

Rheumatoid arthritis is a complex multifactorial disease, and includes a range of immune, neuroendocrine, and psychosocial variables (Glocker et al., 2006). The gene-environment interaction triggers the formation of autoantibodies like rheumatoid factor (RF), anti-citrullinated protein antibodies (ACPA), which leads to the onset and maturation of the disease (Malmström et al., 2017).

Air pollutants such as silica, vehicular smoke, tobacco smoking, etc., lead to an activation of toll-like receptors located in the lung mucosal lining (Bauer et al., 2012; Malmström et al., 2017). Further, a calcium-mediated enzyme peptidyl arginine deiminase (PAD) triggers the process of citrullination.

The binding of ACPAs to surface-expressed citrullinated proteins like Grp78 enhances the activity of nuclear factor kappa B (NFkB) and production of interleukin (IL)-6, and tumor necrosis factor-alpha (TNF-α) (Guo et al., 2018). NFkB is a protein complex that controls transcription of DNA, cytokine production and cell survival. IL-6 acts as pro-inflammatory cytokine and acts as mediator of acute phase response. TNF-α is a cell signaling cytokine involved in systemic inflammation.

The role of pleiotropic cytokine transforming growth factor-beta (TGF-β) is not well established, but the majority of studies found it to be anti-inflammatory and immune-modulatory (Sanjabi et al., 2009). There is an infiltration of leukocytes and pro-inflammatory mediators like IL-17A in the synovial compartment of the affected joint (Firestein and McInnes, 2017). The fulminant disease results in the formation of hyperplastic synovium, damaged cartilage, bony erosions, and extensive systemic inflammation. The essential neuro-endocrine pathway, the hypothalamic-pituitary-adrenal (HPA) axis, modulates the inflammatory response. There is a dysregulation of the HPA axis in RA patients (Cutolo et al., 2002). RA patients show adrenal insufficiency due to which they are unable to mount an appropriately enhanced glucocorticoid response to increased secretion of pro-inflammatory cytokines such as IL-1, IL-6, and TNF-α (Jessop and Harbuz, 2005). There is an insufficient vagal nerve activity in RA, which is the main nerve of the parasympathetic division of the autonomic nervous system and regulates metabolic homeostasis (Koopman et al., 2016, 2017). The vagus nerve stimulation targeting the inflammatory reflex modulates TNF production and reduces inflammation, hence inhibits cytokine production and attenuates disease severity in RA (Koopman et al., 2016). Various genetic, epigenetic, and environmental factors through distinct pathways and neurotransmitters are highly involved in altering the psycho-neuro-immune axis, resulting in the emergence of disease (González-Díaz et al., 2017). Significant stresses from different types of stimuli lead to pro-inflammatory load in RA patients due to defective stress response systems. There is bidirectional interaction between the nervous system and the immune system through soluble cytokines and mediators. The cytokines can traverse the nervous system either by crossing the blood-brain barrier, by diffusion through circumventricular organs or by directly stimulating the peripheral nerves (Otmishi et al., 2008). Stress triggers an aberrant immune response by the release of neuroendocrine hormones leading to precipitation of RA symptoms (Hassett and Clauw, 2010).

Rheumatoid arthritis patients need lifelong medicines to regulate the hyperactive immune system and live under constant anxiety and fear of fatal consequences like irreversible disability. So there is a dire need for its redressal by searching a cost-effective complementary form of treatment with minimal side-effects. Even the type of stressor, the individual's response, and reactivity to stress may prove to be an important prognostic factor in disease onset and progression. Hence, the stress management therapies, along with the traditional pharmacotherapy interventions, may provide a greater sense of control over stress-induced RA symptoms and are capable of producing critical clinical benefits (Tolahunase et al., 2017, 2018; Gautam et al., 2019). Yoga is one such safe and effective therapy to improve physical functions along with stress reduction (Li and Goldsmith, 2012). Various studies have demonstrated the decisive role of yoga and meditation on stress, anxiety, and depression by eliciting a relaxation response that induces temporal transcriptome changes in energy metabolism, insulin secretion and inflammatory pathways (Li and Goldsmith, 2012; Bhasin et al., 2013; Pascoe et al., 2017). Yoga also improves joint flexibility, range of motion, posture, muscle strength, coordination, and thus self-competence and emotional resilience (Polsgrove et al., 2016). Various scientific reports summarized the beneficial effects of yoga on chronic disorders like hypertension, major depressive disorder, diabetes mellitus, aging, insomnia, glaucoma, obesity, and asthma. Yoga aids in the amelioration of mental stress and depression by improving the feeling of perceived well-being (Hagins et al., 2013; Rima et al., 2016; AlKhathami et al., 2017; Tolahunase et al., 2017, 2018; Dada et al., 2018; Sankalp et al., 2018).

Yoga optimizes the levels of oxidative stress as it reduces ROS (reactive oxygen species) levels, reduces 8-OHdG (8-hydroxy-2′-deoxyguanosine) levels, and improves TAC (total antioxidant capacity) levels (Rima et al., 2016; Tolahunase et al., 2017, 2018; Gautam et al., 2019). Yoga also tends to normalize the levels of cortisol, improves brain-derived neurotrophic factor (BDNF), lipid profiles, diabetic profile, and inflammatory processes (Tolahunase et al., 2017, 2018; Raveendran et al., 2018). The pathophysiological association between depression and RA poses a similarity of alteration in biomarkers of both stress response and RA (Margaretten et al., 2011; Matcham et al., 2013).

The role of yoga as an effective intervention has been documented in the literature to assist in the management of chronic diseases like RA concerning its clinical symptoms like pain perception, stress management, disability outcomes, sleep quality, functional ability, QOL and psychosocial outcomes (Evans et al., 2011, 2013; Telles and Singh, 2012; Middleton et al., 2013; Gautam et al., 2019).

Further exploration of a possible mode of action underlying the therapeutic effect of yoga at a cellular level is required and to establish how regular yoga practice affects the systemic biomarkers at the level of the psycho-neuro-immune axis to form a mind-body communication especially, in RA. We hypothesized that lowering of the stress response and elicitation of relaxation response by 8 weeks of a yoga-based lifestyle intervention (YBLI) in active RA patients reduces disease activity by altering the psycho-neuro-immune axis and its associated molecular and genetic markers. Current evidence is limited to establish a link between the multi-faceted dimensions of yoga and how it acts on the dysfunctional immune system and lowers the stress response. With this context in mind, the primary aim of the present study was to investigate the effects of 8-week YBLI on disease activity, psycho-neuro-immune axis markers, gene expression patterns, and QOL. We tested the biomarkers associated with the psycho-neuro-immune axis, including (a) inflammatory markers- IL-6, IL-17A, TNF-α, and TGF-β; (b) mind-body communicative markers- BDNF, DHEAS (dehydroepiandrosterone sulfate), β-endorphin and sirtuins. The expression patterns of the following genes investigated were: *IL-6, TNF-α, NFKB1* (nuclear factor kappa B subunit 1), *TGF-β*, and *CTLA4* (cytotoxic T-lymphocyte-associated protein 4). The inflammatory markers and inflammatory gene expression profile determine the level of systemic as well as local inflammation. The mind-body communicative markers are directly associated with the regulation of neuroplasticity (Tolahunase et al., 2018). In RA, the neuro-immunological mechanisms of immune dysfunction converge in the central nervous system (CNS) to alter molecular programs, neurogenesis, and plasticity and may lead to the development of co-morbid depression. As the immune system is an essential component of the physiological stress-sensing pathways and closely interacts with the body's primary integrative mind-body communicative systems, including the HPA axis, in mutually regulatory feed-forward and feedback loops, so we included these markers.

MATERIALS AND METHODS

Study Design

The design of the present study was a prospective, single-blinded, randomized, controlled trial with active RA patients aimed at investigating the effects of 8 weeks of YBLI on disease activity, psycho-neuro-immune axis markers, gene expression patterns, and QOL. The study protocol complies with the Consolidated Standards of Reporting Trials (CONSORT) guidelines (Figure 1). The study initiated after obtaining ethical clearance (IECPG-211/24.02.2016) from the Institutional Ethics Committee of AIIMS, New Delhi, India, and registration under the clinical trials registry, India (REF/2016/01/010500). All the participants gave written informed consent before the commencement of the study protocol.

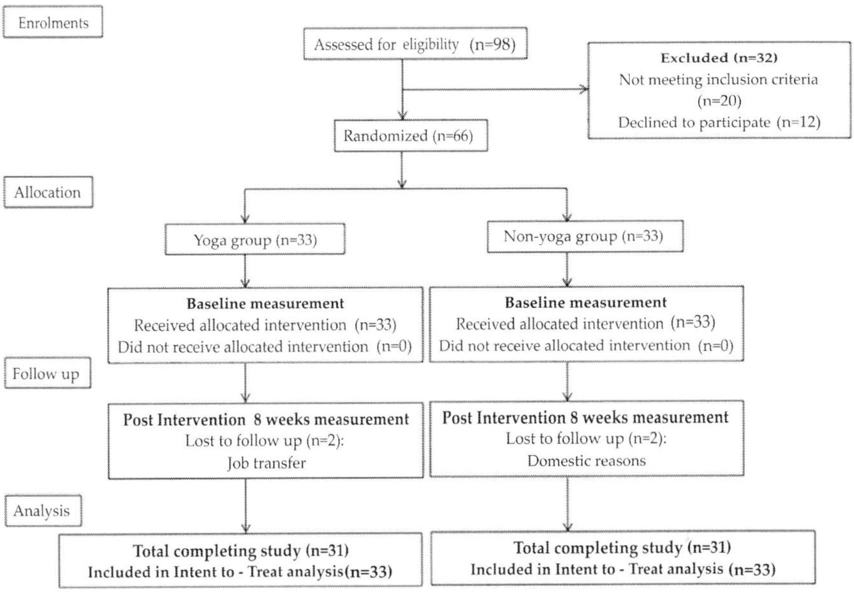

Figure 1. A consort flow diagram of the study.

Participants and Eligibility Criteria

Rheumatoid arthritis patients from the outpatient unit of the Rheumatology department of AIIMS, New Delhi, India, were screened for the following inclusion/exclusion criteria.

Inclusion Criteria for RA Patients

1. Patients of 18–60 years old diagnosed with RA as per 2010 ACR/EULAR RA classification criteria, whose disease activity score 28 erythrocyte sedimentation rate (DAS28-ESR) was >2.6 and were on routine medical treatment for at least 6 months.
2. No participation in a clinical trial within 4 weeks before the study.

Exclusion Criteria for RA Patients

1. Patients with any other autoimmune diseases other than RA or chronic systemic disease that could affect QOL.
2. Any history of administration of oral or intra-articular steroids in previous 6 months, or intake of any form of homeopathic, ayurvedic or herbal mineral supplementation.

3. Those who were already practicing yoga or meditation in any way.
4. Pregnant women and lactating mothers.
5. Patients who were physically unfit for yoga like who underwent any kind of joint replacement surgery, trauma involving the musculoskeletal system like a fracture, etc.
6. Those who were not willing to participate in the study.

Sample Size Calculation

The sample size calculations for the study assumed to detect a standardized effect size [difference in mean change in DAS28-ESR between the two groups/ pooled standard deviation (SD)] of 0.8 with a 95% confidence level and 80% power, considering the mean and SD of a previous study by Evans et al. (2011). The required sample size was 25 in both groups. Considering some loss to follow-up, we enrolled a total of 66 patients in the study and randomized into two groups- yoga group (33 patients) and non-yoga group (33 patients).

Randomization

The sealed opaque envelopes consisted of the computer-generated random numbers, with the assistance of the web tool research randomizer[1], used to randomize the patients into the yoga group or non-yoga group. The participants were not blinded to the study, whereas all investigators who interviewed patients, conducted experiments, and performed statistical analyses blinded to the group status of the patients.

Intervention

A total of 66 participants randomized into two groups - yoga group (n = 33) and non-yoga group (n = 33). The study participants were all asked to undergo a clinical evaluation and provide a blood sample at baseline.

Yoga-Based Lifestyle Intervention (Yoga Group)

Participants of the yoga group were administered a standardized YBLI for 8 weeks, which was suitable for active RA patients in such a way that it did not cause any further irritation to already inflamed joints. The patients with deformed joints and limitations advised undergoing customized physical postures. The comprehensive YBLI program incorporated the components of Patanjali's ashtanga yoga. Briefly, the intervention comprised of yogic practices including asanas (physical postures), pranayama (breathing techniques), dhyana (meditation), and savasana (relaxation techniques) followed by interactive counseling sessions on yoga, stress management, nutrition, as well as personal lifestyle management. YBLI was administered five times a week for 120 min duration per session for 8 weeks (Table 1). The sessions were conducted by the registered and well-qualified yoga instructors at Laboratory for Molecular Reproduction & Genetics, Department of Anatomy, AIIMS, New Delhi, India. As such, there was no home regimen to be followed for the yoga practice due to the extensive time commitment of each session, but the patients were not discouraged for the home practice. The patients were encouraged to incorporate yoga into their lifestyle after the end of the 8 weeks of intervention. The patients of this group continued with their routine DMARDs as per the prescription of rheumatologists.

Table 1. Details of activities in a single day session of the yoga-based lifestyle intervention (YBLI) program.

S. No.	Practice to be done			Duration
1.	Session preparation instructions			2 min
	Starting prayer			5 min
	Yogic sukshma vyama	Finger loosening, wrist loosening, elbow loosening, shoulder loosening, toe bending, ankle bending, knee cap tightening, patella movement, knee bending, and hip rotation		10 min
	Yogic sthool vyama	Rekhagati, Sarvaangpushti		
2.	Yogasana	Standing	Trikonasana	5 min
			Katichakrasana	
			Tadasana	
			Virabhadrasana	
		Sitting	Gomukhasana	5 min
			Paschim-utaanasana	
			Shashaankasana	
			Vakrasana	
		Prone	Ek-pada-shalabhasana	5 min
			Bhujangasana	
			Poorna- shalabhasana	
			Makarasana	
		Supine	Uttanapadasana	5 min
			Setubhandhasana	
			Pavanmuktasana	
			Matsyasana	
3.	Relaxation		Savasana	10 min
4.	Pranayama		Kapalbhati	20 min
			Ujjayi	
			Nadishodhana	
			Bhramari	
5.	Nada anusandhana, AUM- Aumkar recitation			3 min
6.	Dhyana (meditation)			15 min
7.	Shanti mantra- closing prayer			5 min
8.	Interactive session/self-directed learning			30 min
	Total			120 min

Usual Care Control (Non-yoga Group)

Patients assigned to the non-yoga group continued with their on-going usual medical care, which included DMARDs prescribed by the rheumatologists. The patients followed their normal day to day physical activities with no change in their daily routine for 8 weeks.

Outcome Measures

Primary Outcome

The primary outcome was to measure the change in disease activity assessed by DAS28-ESR from baseline (day 0) to 8 weeks (Nishimoto and Takagi, 2010). The composite primary outcome was to evaluate the levels of psycho-neuro-immune axis markers including (a) inflammatory markers- (IL-6, IL-17A, TNF-α, and TGF-β); (b) mind-body communicative markers- BDNF, DHEAS, β-endorphin and sirtuins and the expression patterns of following genes: *IL-6, TNF-*α, *NFKB1, TGF-*β, and *CTLA4.*

Secondary Outcome

The secondary outcome was to measure the change in QOL of RA patients, which was measured by the validated generic WHO Quality of Life BREF (WHOQOL-BREF) questionnaire.

Study Assessments

The assessment of study parameters was at day 0 (baseline) and 8th week (follow-up) of the intervention.

Measurement of Clinical Parameters

DAS28-ESR was used to assess the disease activity of RA patients. The DAS28-ESR consisted of four components: tender joint count, swollen joint count, visual analog scale (VAS) score of the patient's global health, and ESR. In DAS28-ESR, a rating of ≤ 2.6 represents remission, >2.6 to 3.2 represents low disease activity, >3.2 to 5.1 represents moderate disease activity, and >5.1 represents high disease activity.

The assessment of QOL was made by the WHOQOL-BREF questionnaire which contains 26 original items, including 2-items examining an individual's overall perception of the QOL and health; 24 items examining 4-domains (D1, physical; D2, psychological; D3, social; and D4, environmental)

(Gholami et al., 2013). The physical domain has 7-items related to activities of daily living, dependence on medicinal substances and medicinal aids, energy, and fatigue, mobility, pain, and discomfort, sleep and rest and work capacity. The psychological domain has 6-items related to body image and appearance, negative feelings, positive feelings, self-esteem, spirituality, personal beliefs, thinking, memory, learning, and concentration. The social relationship domain has 3-items related to personal relationships, social support, and sexual activity. The environmental domain has 8-items on financial resources, freedom, physical safety, home environment, health and social care and their availability, opportunities for acquiring new information and skills, participation in recreation and leisure activities, physical environment, and transport. The WHOQOL-BREF questionnaire depicts a score ranging from 0 to 100, in which a higher score denotes a better QOL.

Measurement of Biochemical Marker Analysis

For the analysis of biochemical markers, 5 ml of the venous blood sample was collected under aseptic conditions by venipuncture and divided into three parts. From the first part (2 ml), the blood was allowed to clot, and the serum was separated. From the second part (2 ml), plasma was isolated by centrifugation at 2000 × *g* for 15 min at 4°C. RNA was isolated from the remaining third part (1 ml) of the blood. The serum and plasma were stored at −80°C until analyzed. ELISA kits used for the estimation of levels of IL-6 (Gen-Probe, Diaclone Diagnostic, France), IL-17A (Gen-Asia Biotech, China), TNF-α (Gen-Probe, Diaclone Diagnostic, France), BDNF (Raybiotech, Inc), DHEAS (Qayee Bio-Technology), β-endorphin (Phoenix Pharmaceuticals, Inc.) and sirtuin 1 (Qayee Bio-Technology). Serum TGF-β levels were estimated by a magnetic bead-based multiplex assay using Bio-Plex Pro TGF-β Assays (Bio-Rad Laboratories Inc., United States) according to manufacturer's guidelines.

RNA Extraction, cDNA Synthesis, and qPCR

Total RNA was isolated from 1 ml of freshly obtained EDTA blood by TRIzol manual method. Complementary DNA (cDNA) synthesis was done by reverse transcribing 1,000 ng of RNA by using the iScript cDNA synthesis kit (Bio–Rad). The CFX96 real–time system (Bio–Rad, CA, United States) quantified the relative gene expression using Brilliant III Ultra–Fast SYBR Green qPCR Master Mix. The protocol for gene amplification was standardized at 35 cycles. Normalization of the amount of expressed mRNA used two internal housekeeping genes *36B4* and β-*actin*. The relative fold of gene expression was done by the $2^{-\Delta\Delta Ct}$ method. Each cDNA product formed was tested in triplicate. The primer sequences were as follows:

*IL-6*_Forward, 5′-GGCACTGGCAGAAAACAACC-3′;

*IL-6*_Reverse, 5′-GCAAGTCTCCTCATTGAATCC-3′;

TNF-α_Forward, 5′-ACCTCCGAGATGACACCATCA-3′;

TNF-α_Reverse, 5′-GGCACTCTGGCACATATTCAC-3′;

*NFKBI*_Forward, 5′-CTGAGGCACTTCTGGGAGC-3′;

*NFKBI*_Reverse, 5′-CTCGAAAGTCTCGGAGCT-3′;

TGF-β_Forward, 5′-GAAGGGAGACAATCGCTTTAGC-3′;

TGF-β_Reverse, 5′-TGTAGACTCCTTCCCGGTTGAG-3′;

*CTLA4*_Forward, 5′-GCTTTCTCCTCACAGCTGT-3′;

*CTLA4*_Reverse, 5′-TTTTCACATTCTGGCTCTGTT-3′;

*36B4*_Forward, 5′-AACATGCTCAACATCTCCCC-3′;

*36B4*_Reverse, 5′-CCGACTCCTCCGACTCTTC-3′;

β-*actin*_Forward, 5′-TGAGAGGGAAATCGTGCGTG-3′;

β-*actin*_Reverse, 5′-TGCTTGCTGATCCACATCTGC-3′.

Statistical Analysis

All statistical analyses carried out on an intent-to-treat basis with the baseline observation carried forward approach using IBM SPSS Statistics for Macintosh, Version 25.0. (IBM Corp. Armonk, NY, United States) and GraphPad Prism (Version 6.01). Chi-square test and Fisher's exact test compared the baseline characteristics between the two groups. The assessment of interaction effects among baseline parameters was carried out by mixed factorial design ANOVA (Peterson et al., 2017). For within-group analysis, paired *t*-test was used to study the difference between pre- to post-intervention for normally distributed data, or Wilcoxon signed-rank tests for continuous variables without normal distribution. For between-group analysis, the repeated measure ANOVA was used to study the intervention effects along with the interaction of time and group. The correlation matrix representing correlation coefficients between variables was calculated based on the Pearson correlation. In order to conduct the mediation analysis, the multiple regression analysis was used to determine the change in which variables significantly explained the associations with the dependent variable, i.e., change in pre and post intervention DAS28-ESR. A p-value < 0.05 was considered as statistically significant.

RESULTS

Participants' Baseline Characteristics

A total of 98 patients were screened for eligibility, out of which 66 were randomized into two groups (each group n = 33) (Figure 1). Baseline demographic characteristics, presenting symptoms, disease activity and QOL of all randomized participants are shown in Table 2. There were no significant differences in the baseline characteristics between the two intervention groups.

Table 2. Baseline characteristics.

Variable	Group		χ^2 value	*t*-value	*p*-value
	Yoga (*n* = 33)	**Non-yoga (*n* = 33)**			
Demographic characteristics					
Age (years)	45.1 (8.7)	43.4 (9.3)	–	0.7484	0.4569
Sex					
Male	5 (15.2)	8 (24.2)	0.8621	–	0.5372
Female	28 (84.8)	25 (75.8)		–	
BMI (kg/m²)	25.9 (5.3)	24.6 (3.4)	–	1.126	0.2785
Kuppuswamy's Socioeconomic Status Scale					
Upper	5 (16.0)	1 (3.2)	4.364	–	0.3590
Upper middle	10 (32.3)	12 (38.8)		–	
Lower middle	5 (9.7)	7 (22.5)		–	
Upper lower	10 (32.3)	12 (32.3)		–	
Lower	3 (9.7)	1 (3.2)		–	
Presenting symptoms					
Early morning stiffness (minutes)	22.4 (3.8)	23.5 (3.7)	–	0.1979	0.8437
Tender joint count (TJC)	6.0 (0.8)	6.4 (0.7)	–	0.4132	0.5143
Swollen joint count (SJC)	4.0 (0.7)	3.7 (0.5)	–	0.2668	0.6927

Disease activity					
DAS28-ESR	5.2 (1.1)	5.0 (0.8)	–	0.8603	0.3928
Quality of life					
WHOQOL-BREF Domain 1	43.3 (10.5)	44.8 (9.9)	–	0.6089	0.5447
WHOQOL-BREF Domain 2	47.2 (8.5)	48.0 (8.3)	–	0.4078	0.6848
WHOQOL-BREF Domain 3	49.0 (6.5)	49.2 (6.7)	–	0.1117	0.9114
WHOQOL-BREF Domain 4	53.5 (8.4)	53.7 (8.6)	–	0.0874	0.9306

Post-intervention Differences in Psycho-Neuro-Immune Axis Markers

As shown in Table 3, there was a significant change observed in the primary outcome, i.e., DAS28-ESR after the 8 weeks of the trial. There was a significant decrease found in DAS28-ESR (−0.8 ± 0.5; $p < 0.0001$ vs. −0.1 ± 0.6; $p = 0.303$) in yoga group as compared to non-yoga group after 8 weeks of intervention. The between group analysis showed a mean difference in DAS28-ESR [0.4, 95% CI (0.3 to 0.6)] with a significant interaction of group and time ($\eta_p^2 = 0.246$; $p \leq 0.001$). A significant decline was noted in the levels of pro-inflammatory cytokine IL-6 [0.5, 95% CI (0.2 to 0.8); $\eta_p^2 = 0.137$; $p = 0.002$], IL-17A [0.1, 95% CI (−5.7 to 6.1); $\eta_p^2 = 0.388$; $p \leq 0.001$] and TNF-α [1.1, 95% CI (0.3 to 1.9); $\eta_p^2 = 0.311$; $p \leq 0.001$] in yoga group after 8 weeks, whereas the levels of anti-inflammatory cytokine TGF-β have shown a significant elevation [−3.3, 95% CI (−6.7 to −0.2); $\eta_p^2 = 0.166$; $p = 0.001$] after 8 weeks of YBLI. There was an upregulation in mind-body communicative markers with a significant mean difference with group and time interaction in BDNF [−2.0, 95% CI (−2.9 to −1.1); $\eta_p^2 = 0.202$; $p < 0.001$], DHEAS [−12.4, 95% CI (−14.5 to −10.3); $\eta_p^2 = 0.602$; $p < 0.001$], β-endorphin [−0.4, 95% CI (−0.7 to −0.8); $\eta_p^2 = 0.431$; $p < 0.001$] and sirtuins [−1.9, 95% CI (−2.9 to −0.9); $\eta_p^2 = 0.362$; $p < 0.001$] after 8 weeks of YBLI. Also, the change in mean within non-yoga group showed significantly increased levels of IL-17A (18.7 ± 23; $p < 0.0001$) and reduced levels of sirtuin 1 (−1.1 ± 2.2; $p < 0.0001$) and β endorphins (−0.7 ± 0.7; $p < 0.0001$) compared to baseline levels.

Table 3. Intent-to-treat analysis: means (SD) and results of within-group and between-group analysis of study outcomes (n = 66; yoga, 33; non-yoga, 33).

Outcome	Yoga group			Non-yoga group			Between groups	Variance analysis/effects			
	Pre	Post	Δ Mean within-group	Pre	Post	Δ Mean within-group	Mean difference (95% CI)[b]	Time		Group × time	
								p^b-value	η2p	p^b-value	η2p
Disease activity score											
DAS28-ESR	5.2 ± 1.1	4.5 ± 0.9	−0.8 ± 0.5	5.02 ± 0.8	4.9 ± 0.8	−0.1 ± 0.6	0.4 (0.3–0.6)	<0.001	0.372	<0.001	0.246
Inflammatory markers											
IL-6 (pg/ml)	3.5 ± 1.3	2.4 ± 1.7	−1.0 ± 1.1	3.8 ± 1.1	3.7 ± 1.5	−0.1 ± 1.4	0.5 (0.2–0.8)	0.001	0.153	0.002	0.137
IL-17A (pg/ml)	195.0 ± 31.0	176.6 ± 30.7	−19.0 ± 25.1	188.4 ± 38.6	207.2 ± 42.2	18.7 ± 23	0.1 (-5.7 – 6.1)	0.962	<0.001	<0.001	0.388
TNF-α (pg/ml)	17.0 ± 3.5	13.7 ± 3.8	−3.3 ± 3.0	17.5 ± 6.9	18.5 ± 5.4	1.0 ± 3.4	1.1 (0.3–1.9)	0.006	0.113	<0.001	0.311
TGF-β (ng/ml)	41.9 ± 16.8	51.3 ± 15.5	93.9 ± 13.6	44.4 ± 14.4	41.8 ± 15.1	−26.2 ± 13.7	−3.3 (-6.7 – 0.2)	0.049	0.059	0.001	0.166
Mind-body communicative markers											
BDNF (ng/ml)	13.3 ± 3.9	17.2 ± 6.1	3.8 ± 4.5	13.5 ± 4.3	13.8 ± 4.7	0.2 ± 2.4	−2.0 (-2.9 – -1.1)	<0.001	0.247	<0.001	0.202
DHEAS (ng/ml)	37.4 ± 5.7	60.3 ± 8.6	22.8 ± 10.6	35.3 ± 4.1	37.3 ± 5.3	2.0 ± 5.9	−12.4 (-14.5 – -10.3)	<0.001	0.683	<0.001	0.602
β-endo (ng/ml)	3.7 ± 1.3	5.3 ± 1.7	1.5 ± 1.7	3.8 ± 0.6	3.1 ± 0.7	−0.7 ± 0.7	−0.4 (-0.7 – -0.8)	0.015	0.089	<0.001	0.431
Sirtuin 1 (ng/ml)	31.8 ± 1.8	36.8 ± 5.9	5.0 ± 5.3	30.7 ± 2.4	29.6 ± 2.4	−1.1 ± 2.2	−1.9 (-2.9 – -0.9)	<0.001	0.187	<0.001	0.362
WHOQOL-BREF score											
D1	43.3 ± 10.3	56.4 ± 10.5	13.1 ± 7.8	44.8 ± 9.9	43.3 ± 7.3	−1.4 ± 7.2	−5.8 (-7.6 – -3.9)	<0.001	0.384	<0.001	0.493
D2	47.2 ± 8.5	60 ± 5.4	12.8 ± 7.4	48.0 ± 8.3	45 ± 9.9	−3.0 ± 6.8	−4.9 (-6.6 – -3.1)	<0.001	0.329	<0.001	0.561
D3	49.0 ± 6.5	53.5 ± 6.7	4.5 ± 4.5	49.2 ± 6.6	46.7 ± 9.1	−2.5 ± 5.4	−1.0 (-2.2 – 0.2)	0.113	0.039	<0.001	0.333
D4	53.5 ± 8.3	54.3 ± 7.3	0.7 ± 3.8	53.7 ± 8.5	52.7 ± 7.6	−0.9 ± 3.8	0.12 (-0.8 – 1.1)	0.799	0.001	0.078	0.048

Δ mean within-group, mean score change within the group; p[b], p-value with Bonferroni correction; η_p^2, partial eta square value representing the effect size; WHOQOL-BREF Domains, D1, physical; D2, psychological; D3, social; and D4, environmental.

Post-intervention Differences in Quality of Life

The change in mean scores of QOL for physical domain D1 (13.12 ± 7.8 vs. -1.4 ± 7.2; $p < 0.0001$), psychological domain D2 (12.8 ± 7.4 vs. -3.0 ± 6.8; $p < 0.0001$), social domain D3 (4.5 ± 4.6 vs. -2.5 ± 5.4; $p < 0.0001$) were found to be statistically significant in yoga group over non-yoga group. A trend toward improvement was noted in environmental domain (domain D4) scores (0.72 ± 3.8 vs. -0.96 ± 3.8; $p = 0.078$) in yoga group as compared to the control group (Figure 2 and Table 3). The variance analysis showed a significant mean score difference in QOL for physical domain D1 [−5.8, 95% CI (−7.6 to −3.9); $\eta_p^2 = 0.493$; $p < 0.001$], psychological domain D2 [−4.9, 95% CI (−6.6 to −3.1); $\eta_p^2 = 0.561$; $p < 0.001$], social domain D3 [−1.0, 95% CI (−2.2 to 0.2); $\eta_p^2 = 0.333$; $p < 0.001$] in yoga group as compared to non-yoga group. The environmental domain D4 did not show any statistically significant mean score difference in both the groups [0.12, 95% CI (−0.8 −1.1); $\eta_p^2 = 0.048$; $p = 0.078$] (Table 3).

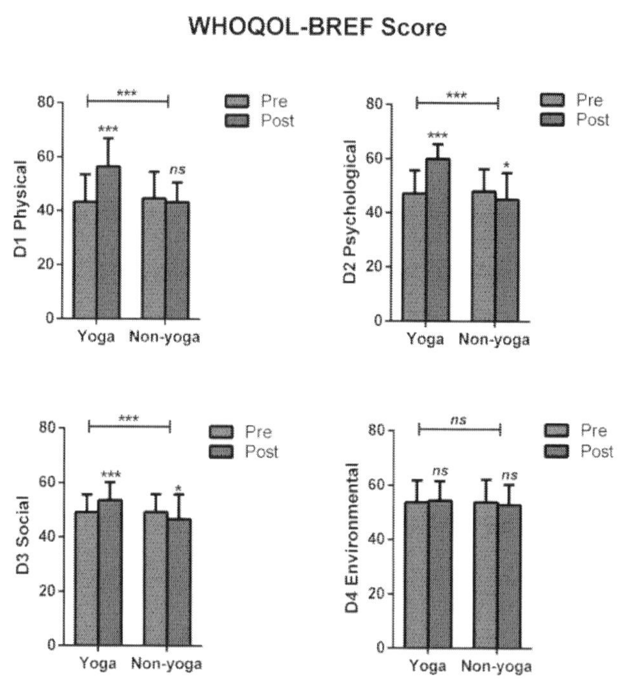

Figure 2. Between and within-group analysis of WHOQOL-BREF Scores in yoga group and non-yoga group. p-value (ns = $p > 0.05$; *$p \leq 0.05$; **$p \leq 0.01$; ***$p \leq 0.001$).

The data showed significant improvement in the overall QOL (item 1) from baseline to week 8, as assessed by the WHOQOL-BREF questionnaire. At the end of the study, no participant had a poor perception of overall QOL after the YBLI, and most of the participants perceived an overall good QOL. This shift in scores from baseline was statistically significant in the yoga group ($p = 0.0007$) (Table 4 and Figure 3). In-line, the results for the overall quality of health (item 2) showed that none of the participants was very dissatisfied after the YBLI, and most participants were satisfied. This shift in scores from baseline was statistically significant in the yoga group ($p < 0.0001$) (Table 5 and Figure 3).

Table 4. Pre–post comparison of World Health Organization quality of life-BREF responses: overall perception of quality of life (item 1) in RA patients.

Variable	**Pre-intervention (baseline)**	**Post-intervention (8th week)**					***p*-values**
		Very poor	**Poor**	**Neither poor nor good**	**Good**	**Very good**	
Yoga group							
Very poor	3	0	1	2	0	0	**0.0007**
Poor	9	0	0	5	3	1	
Neither poor nor good	13	0	0	2	9	2	
Good	8	0	0	0	5	3	
Very good	0	0	0	0	0	0	
Non-yoga group							
Very poor	3	0	3	0	0	0	**0.7084**
Poor	9	1	6	2	0	0	
Neither poor nor good	12	0	1	8	3	0	
Good	9	0	0	1	7	1	
Very good	0	0	0	0	0	0	

Figure 3. The improvement in overall quality of life and overall quality of health after yoga-based lifestyle intervention.

Table 5. Pre–post comparison of World Health Organization quality of life-BREF responses: Overall perception of quality of health (item 2) in RA patients.

Variable	Pre-intervention (baseline)	Post-intervention (8th week)					*p*-values
		Very dissatisfied	Dissatisfied	Neither dissatisfied nor satisfied	Satisfied	Very satisfied	
Yoga group							
Very dissatisfied	4	0	2	1	1	0	**<0.0001**
Dissatisfied	10	0	1	3	5	1	
Neither dissatisfied nor satisfied	18	0	0	4	14	0	
Satisfied	1	0	0	0	0	1	
Very satisfied	0	0	0	0	0	0	
Non-yoga group							

Very dis-satisfied	2	1	0	1	0	0	**0.1775**
Dissatisfied	17	2	7	5	3	0	
Neither dis-satisfied nor satisfied	11	0	2	6	3	0	
Satisfied	3	0	0	0	1	2	
Very satis-fied	0	0	0	0	0	0	

Gene Expression Analysis

The results showed significant down regulation in relative mRNA expression levels of *IL-6* ($p = 0.001$), *TNF-α* ($p = 0.0271$) and *CTLA-4* ($p = 0.0129$) in the yoga group as compared to the non-yoga group (Figure 4). The mean axis fold change of these transcripts was as follows in the yoga vs. non-yoga group: *IL-6* (−2.6 vs. 0.4), *TNF-α* (−2.7 vs. 1.35), and *CTLA-4* (−0.5 vs. 2.2), respectively. The mRNA expression levels of *NFKB1* were not found to be different statistically ($p = 0.9137$) with a mean axis fold change of 0.6 vs. 1.1 in the yoga vs. non-yoga group (Figure 4). In the yoga group, the mRNA expression levels of *TGF-β* ($p = 0.0052$) were significantly upregulated with the mean axis fold change of 5.5 vs. 3.8 in the yoga vs. non-yoga group (Figure 4).

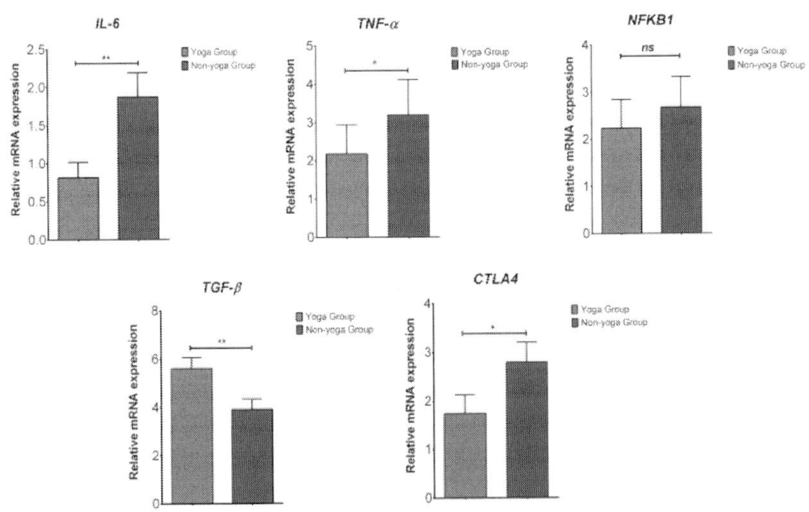

Figure 4. The relative m RNA expression levels of dysregulated transcripts in yoga group and non-yoga group. *IL-6* (interleukin-6), *TNF-α* (tumor necrosis factor alpha), *NFKB1* (nuclear factor kappa B subunit 1), *TGF-β* (transforming

growth factor beta), and *CTLA4* (cytotoxic T-lymphocyte-associated protein 4); p value (ns = $p > 0.05$; *$p \leq 0.05$; **$p \leq 0.01$; ***$p \leq 0.001$).

Group × Gender Interactions

There was no significant difference in mean DAS28-ESR values between men and women at baseline (DAS28-ESR values 5.1 ± 0.1 and 5.2 ± 0.3 mean ± SD, respectively). Separate analyses for males and females were performed to overcome baseline violations in disease activity and to explore further specific gender effects. Interaction effects including group and gender indicated differential responses to YBLI for women for DAS28-ESR [$F(1,27) = 70.3$; $p < 0.001$], BDNF [$F(1,27) = 22.6$; $p < 0.001$], sirtuins [$F(1,27) = 24.3$ $p < 0.001$], IL-17A [$F(1,27) = 15.2$; $p = 0.001$], IL-6 [$F(1,27) = 23.2$; $p < 0.001$], TNF-α [$F(1,27) = 38.5$; $p < 0.001$], TGF-β [$F(1,27) = 10.3$; $p = 0.003$]. Clinical improvement was more significant for the women in yoga group [mean between-group difference of change [95% confidence interval (CI)]: female = 0.6, 95% CI (0.3 to 0.9) $p = 0.0003$; male = 0.8, 95% CI (0.04 to 1.7) $p = 0.0407$ (Figure 5).

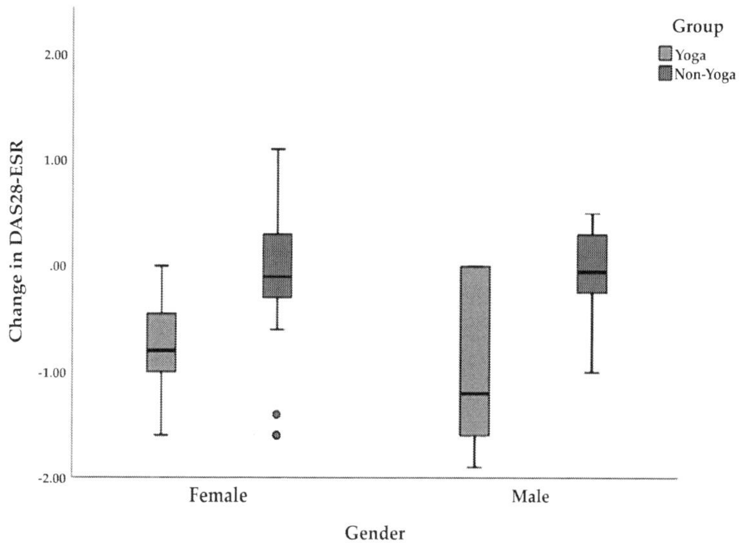

Figure 5. Gender interactions for change in disease activity after the intervention. Mean change of DAS28-ESR score with 95% CI for males in the yoga and the control groups; $p = 0.0407$ for between-group differences of change in the study, adjusted for baseline value. Mean change of DAS28-ESR score with 95% CI for females in the yoga and the control groups; $p = 0.0003$ for the between-group difference of change on the study, adjusted for baseline value.

Correlation between Psycho-Neuro-Immune Markers, Gene Expression and Quality of Life

The analysis showed a correlation between the pre and post-intervention change among the tested variables (Table 6). The improvement of DAS28-ESR was correlated positively with the changes of WHOQOL-BREF scores and mind-body communicative markers (BDNF, DHEAS, β-endorphin, and sirtuin) but negatively with the changes of inflammatory markers (IL-6, IL-17A, TNF-α). However, both the groups showed non-significant relationships with the pre and post intervention change of different variables with respect to change in DAS28-ESR (Table 7). The fact that the observed *p*-values did not fall below the established alpha level of 0.05 which indicated that the association among the measured variables not significantly affected the dependent variable by the inclusion of the mediators in the model; in other words, there is no evidence of mediation.

Table 6. Correlations analysis of pre-post differences (changes) between primary and secondary outcomes (n = 66)..

		DAS28-ESR	Inflammatory markers				Mind-body communicative markers				WHOQOL-BREF				Axis fold change				
		DAS28-ESR	Inflammatory markers				Mind-body markers	communicative			WHOQOL-BREF				Axis fold change				
		DAS	IL-6	IL-17A	TNF-α	TGF-β	BDNF	DHEAS	β-Endo	Sirtuin1	D1	D2	D3	D4	IL6	TNF-α	NFKB1	TGF-β	CTLA4
DAS28-ESR	DAS																		
Inflammatory markers	IL-6	0.4**																	
	IL-17A	0.3*	0.3*																
	TNF-α	0.3*	0.4**	0.4**															
	TGF-β	-0.2	-0.3*	-0.4**	-0.3**														
Mind-body communicative markers	BDNF	-0.4**	-0.4**	-0.2	-0.1	0.3**													
	DHEAS	-0.5**	-0.3**	-0.5**	-0.6**	0.4**	0.3**												
	β-Endo	-0.4**	-0.3*	-0.4**	-0.5**	0.1	0.3*	0.5**											
	Sirtuin1	-0.4**	-0.4**	-0.5**	-0.5**	0.5**	0.4**	0.5**	0.6**										
WHOQOL-BREF	D1	-0.4**	-0.2	-0.5**	-0.5**	0.3*	0.3*	0.7**	0.4**	0.4**									
	D2	-0.5**	-0.5**	-0.6**	-0.5**	0.4**	0.4**	0.8**	0.5**	0.5**	0.7**								
	D3	-0.3*	-0.2	-0.4**	-0.3*	0.2	0.2	0.6**	0.4**	0.3*	0.5**	0.4**							
	D4	-0.3*	-0.2	0	-0.3**	0	0	0.3*	0.2	0.2	0.2	0.1	0.2						
Axis fold change	*IL-6*	0.3*	0.2	0.4**	0.3*	-0.2	0	-0.4**	-0.3**	-0.2	-0.1	-0.2	-0.3*	0					
	TNF-α	0	0	0.1	0.2	-0.2	-0.1	-0.1	-0.2	-0.1	0	-0.1	-0.1	0.1	0.2				
	NFKB1	0	-0.1	0.1	-0.1	0.3*	0.1	0.1	-0.1	0.1	0.1	0	0.1	0.1	0.3**	0.3*			
	TGF-β	-0.1	-0.2	-0.3*	-0.2	0.2	0.3**	0.3*	0.1	0.1	0.4**	0.3*	0.2	0.1	0.1	0	0.4**		
	CTLA4	0.2	0.2	0.2	0.3*	-0.2	-0.2	-0.2	-0.4**	-0.3**	-0.2	-0.3**	-0.2	0	0.4**	0.2	0.1	0.1	

Color Key *Lowest* → *Highest* (-1, 0, +1)

*Correlations were significant at the level of p < **0.01 and p < *0.05 (2-tailed).*

Table 7. Predictors of change in DAS28-ESR with group interactions.

Group	Variable	Unstandardized co-efficient		Standardized co-efficient	*t*	*p*-value
		B	Standard error	β		
Yoga						
	Constant	–0.261	0.286		–0.911	0.373
Inflammatory markers						
	Δ IL-6	0.036	0.120	0.073	0.305	0.764

	Δ IL-17A	0.004	0.005	0.176	0.803	0.431
	Δ TNF-α	–0.071	0.040	–0.392	–1.769	0.092
	Δ TGF-β	1.476*E*–06	0.000	0.037	0.158	0.876
Mind-body communicative markers						
	Δ BDNF	–0.025	0.028	–0.209	–0.892	0.383
	Δ DHEAS	–0.013	0.011	–0.264	–1.277	0.216
	Δβ-*endo*	–0.063	0.066	–0.204	–0.956	0.350
	Δ Sirtuin 1	–0.014	0.026	–0.140	–0.546	0.591
WHOQOL-BREF score						
	Δ D1	–0.040	0.021	–0.572	–1.902	0.072
	Δ D2	0.034	0.022	0.464	1.576	0.131
	Δ D3	0.007	0.026	0.059	0.267	0.792
	Δ D4	–0.032	0.029	–0.232	–1.125	0.274
Non-yoga						
	Constant	–0.071	0.224		–0.317	0.754
Inflammatory markers						
	Δ IL-6	0.035	0.136	0.078	0.258	0.799
	Δ IL-17A	0.000	0.006	0.007	0.030	0.977
	Δ TNF-α	0.038	0.070	0.203	0.539	0.596
	Δ TGF-β	6.091*E*–06	0.000	0.132	0.446	0.661
Mind-body communicative markers						
	Δ BDNF	–0.091	0.085	–0.348	–1.070	0.297
	Δ DHEAS	–0.006	0.023	–0.059	–0.267	0.792
	Δβ-*endo*	0.082	0.196	0.098	0.420	0.679
	Δ Sirtuin 1	0.014	0.066	0.049	0.215	0.832
WHOQOL-BREF score						
	Δ D1	0.012	0.020	0.133	0.572	0.574
	Δ D2	–0.024	0.023	–0.259	–1.055	0.304
	Δ D3	0.017	0.025	0.143	0.658	0.518
	Δ D4	–0.032	0.036	–0.190	–0.872	0.393

DISCUSSION

In the present study, we found that 8 weeks of YBLI significantly reduced disease activity, normalized the biomarkers of the psycho-neuro-immune axis with associated changes in gene expression patterns, and improved QOL. With a decline in DAS28-ESR, we observed reduced levels of inflammatory markers, elevated mind-body communicative markers associated with the psycho-neuro-immune axis in the yoga group. The yoga group showed the downregulation of *IL-6, TNF*-α, and *CTLA4* and upregulation of *TGF*-β.

Our study also highlights a significant improvement in three of four domains of QOL after 8 weeks of YBLI. There was an improvement in various molecular and genetic markers associated with the psycho-neuro-immune axis. Also, there was a favorable clinical outcome of RA with a reduction in disease activity and improvement in QOL. These findings suggest that yoga may induce molecular remission in RA by influencing its pathobiology by targeting mind-body communications.

In the present study, the impact of YBLI was beneficial as there was a significant reduction in pro-inflammatory markers (IL-6, IL-17A, and TNF-α) and elevation in anti-inflammatory markers (TGF-β). Anti-inflammatory effects of yoga try to re-establish optimal immune-homeostasis and bring about natural immunological tolerance to overcome the autoimmune conditions (Gautam et al., 2019). Our results are in concordance with other studies that demonstrated yoga's potential role in reducing sympathetic over-activity, increasing vagal activity, and hence lowering inflammation (Rajbhoj et al., 2015; Vijayaraghava et al., 2015; Shete et al., 2017). Yoga and meditation have various physiological implications that help in reducing psychological stress parameters, which help in healthy aging and cellular longevity (Tolahunase et al., 2017, 2018). Yoga intervention leads to reduced levels of oxidative stress and increased telomerase activity in apparently healthy individuals (Tolahunase et al., 2017). The long term practice of meditation exhibited downregulation of T cell receptor signaling and inflammatory response pathways associated with *NFKB, RELA, TNFR2* transcription factors (Bhasin et al., 2013). *NFKB* is a potential bridge between inflammation, psychosocial stress, aging, oxidative cellular activation (Lawrence, 2009). Our results show a non-significant decline in relative mRNA expression of *NFKB1* in the yoga group as compared to the non-yoga group after a short-term intervention of 8 weeks. However, we observed a declining trend of *NFKB1* expression in the yoga group as compared to the non-yoga group.

IL-6 is a pleiotropic cytokine encoded by the *IL-6* gene located on chromosome 7, plays a crucial role in the pathogenesis of RA, and has a strong positive correlation with the disease activity and joint destruction. It mediates its pro-inflammatory actions by binding to its transmembrane or soluble receptors and further activating gp130, a transmembrane protein, to initiate the IL-6 signaling pathway. IL-6 signaling also activates various transcriptional factors via the JAK-STAT signaling pathway, which leads to proliferation, differentiation, and activation of CD4+ T cells (Su et al., 2017). IL-6 in the presence of TGF-β also promotes the differentiation of Th17 cells, which secrete IL-17A, another potent pro-inflammatory cytokine. At the same time, it inhibits TGF-β induced differentiation of regulatory T (T reg) cells (Martinez et al., 2008). Hence, IL-6 creates an imbalance of immune-homeostasis by increasing the Th17 population over T reg cell populations. IL-6 is also a potent activator of the HPA axis, induces acute phase reactions, and causes the production of autoantibodies by stimulating B cells (Venihaki et al., 2001). In RA, there is an autonomic dysfunction which is represented by a low parasympathetic activity. The vagus nerve plays an important role in regulating inflammation as it decreases production of pro-inflammatory cytokines such as IL-6, TNF-α, etc. and inhibits the migration of immune cells to sites of inflammation (Bonaz et al., 2017). Heart rate variability (HRV) is a reliable index of cardiac-vagal regulation (Breit et al., 2018). HRV is inversely related to the levels of inflammatory markers (Bonaz et al., 2017; Breit et al., 2018). The vagus nerve stimulates the HPA axis through its afferent fibers to release glucocorticoids by the adrenal glands and is also involved in the cholinergic anti-inflammatory pathway through a vago-vagal reflex. The sympathetic nervous system and the vagus nerve interact both through a vago-sympathetic pathway involving vagal afferent fibers and a vago-splenic pathway through vagal efferent fibers (Cooper et al., 2015). The modulation of vagus nerve is able to improve various inflammatory disorders, like irritable bowel syndrome, inflammatory bowel diseases, RA, obesity, and pain, etc. which is possible via pharmacological manipulation, vagus nerve stimulation, nutritional therapies, yoga intervention and physical exercise (Bonaz et al., 2017). In inflammatory conditions, the HPA axis is continuously under stress. Stressors/pro-inflammatory cytokines trigger the hypothalamus to release corticotrophin-releasing hormone (CRH). CRH, in turn, acts on anterior pituitary and induces the release of adrenocorticotropic hormone (ACTH), which in turn stimulates the adrenal gland to produce and release cortisol. Cortisol counteracts the stressor and reduces the existing inflammation (Silverman et al., 2005). But due to continuous inflammatory stimuli, the HPA axis is sensitized, and there is an initiation of negative feedback regulation of cortisol on the hypothalamus and anterior pituitary. This condition leads to adrenal insufficiency. Various studies have shown that yoga modulates stress and inflammation by suppressing hypersensitive HPA axis and modulating cortisol levels (Arora and Bhattacharjee, 2008).

Rheumatoid arthritis patients experience clinical symptoms like joint stiffness, pain, and functional disability mostly in the early morning as these symptoms closely follow the circadian rhythm of the pro-inflammatory cytokines (Sierakowski and Cutolo, 2011). Human pro-inflammatory cytokine production also exhibits diurnal rhythmicity. When the levels of pro-inflammatory cytokines are highest during the night, and early morning, there exists an imbalance in serum cortisol level (lowest) and serum melatonin level (highest). There is an increase in nocturnal anti-inflammatory cortisol secretion, which is insufficient to suppress the ongoing inflammation, resulting in the morning symptoms characteristic of RA. This imbalance of anti-inflammatory effects exerted by cortisol and pro-inflammatory effects by melatonin plays a crucial role in the pathogenesis of RA (Cutolo et al., 2003; Cutolo, 2019). A low dosage of corticosteroid replacement therapy aids in relieving aggravated clinical symptoms. Also, melatonin inhibitors may prove to be therapeutic for RA in relieving the symptoms associated with RA (Cutolo et al., 2003). The previous study from our laboratory demonstrated a significant decline in the cortisol levels after 12 weeks of YBLI in a healthy population as well as in depression patients (Tolahunase et al., 2017, 2018). The therapeutic role of yoga and the time of its administration during the day become relevant to study its impact on circadian rhythm in RA patients. Our study also demonstrates that yoga normalizes the circulating levels (IL-6, IL-17A, TNF-α) and mRNA transcript levels of pro-inflammatory cytokines (*IL-6, TNF*-α), hence reduces systemic and local inflammation in RA. Studies suggest that the vagus nerve has a profound anti-inflammatory effect and its activity is increased by yoga as indexed by HRV. Yoga also brings a balance between sympathetic and parasympathetic limbs in inflammatory conditions and restores autonomic reflex regulatory mechanisms.

TGF-β is another important pleiotropic cytokine which by its regulatory functions, helps to maintain immune-homeostasis and induces peripheral tolerance. TGF-β supports the survival of naturally occurring T reg cells and promotes the differentiation of induced T reg cells with IL-2 and retinoic acid. The protective nature of T reg cells keeps a track on the hyperactive immune system and prevent autoimmunity (Gonzalo-Gil and Galindo-Izquierdo, 2014). The results from our study showed that yoga holds the immune-regulatory potential and reduces the severity of RA by re-establishing immunological tolerance. CTLA4 is a co-inhibitory molecule that modulates the CD28: CD80/86 co-stimulation signal, thus inhibits T cell response. CTLA4 has a stronger affinity to bind with CD80 and CD86 and sends inhibitory signals to T cells. CTLA4 is also found in T regs and contributes to their inhibitory function. The blockade of CTLA-4 reduces the process of angiogenesis, suppresses *NFKB* activation, and promotes osteoclastogenesis (Cutolo et al., 2016). Our results have shown that the yoga group showed significant downregulation of mRNA expression levels of *CTLA4* as compared to the non-yoga group, further confirming the immune-regulatory role of YBLI.

We have also seen a marked improvement in the mind-body communicative markers, which is indicated by increased levels of BDNF, DHEAS, β endorphins, and sirtuin-1, followed by 8 weeks of YBLI in RA patients. YBLI elevates the levels of BDNF, a cardinal biomarker of neuroplasticity. BDNF possesses neurotrophic and neuroprotective abilities (Naveen et al., 2016; Gagrani et al., 2018). Our study confirms that YBLI increases BDNF and improves mind-body communications that regulate neuroplasticity and reduce associated stress cascade. DHEAS has neuroprotective, antioxidant, and anti-inflammatory properties. There is dysregulation of DHEAS levels in RA patients (Maninger et al., 2009). Previous studies have shown that DHEAS levels are upregulated in RA patients when administered with TNF antagonists and improved adrenal functioning (Schmidt et al., 2005; Maninger et al., 2009). There is a significant elevation in the DHEAS levels after 8 weeks of YBLI, which helps in the maintenance of emotional regulation, neurocircuits, and memory. Upregulated DHEAS levels after yoga intervention also reduces depression severity, as seen in primary depression patients (Tolahunase et al., 2018). β endorphins are endogenous morphine secreted by the pituitary and reached all the tissues via diffusion. This neuro-hormone acts as a neuro-regulator and circulating levels of which tend to increase with a particular form of physical activity of sufficient intensity and duration (Schwarz and Kindermann, 1992). Hence, yoga also possesses the potential to improve the levels of β endorphins and develops a positive feeling with a sense of happiness, well-being, and security in already depressed RA patients. Yoga also resulted in the upregulation of sirtuins, which promote cellular longevity by deactivation of p53 mediated pathways that are involved in apoptosis, growth arrest, or senescence in response to cellular stress (Tolahunase et al., 2018; Gautam et al., 2019). Our findings suggest that yoga has the ability to reorganize the structure and function of nervous system which was manifested at genetic and biochemical levels by downregulation of inflammatory markers, upregulation of markers of neuroplasticity and reduction in DAS28-ESR scores. In our study, it was found that clinical improvement was more significant for the women in yoga group in terms of DAS28-ESR and various other markers, i.e., BDNF, sirtuins, IL-17A, IL-6, TNF-α, and TGF-β. Women tended to show higher reductions in disease activity and depression severity. This may be due to transitory and fluctuating sex hormone levels that induce the continuous functional adaptation of the CNS throughout the lifespan, particularly in females (Juraska et al., 2013). It has also been observed that neurogenesis is higher in female than in males, probably because of the variations in gonadal hormones as estrogens modulate spatial and contextual memory in females (Duarte-Guterman et al., 2015). The results of this study also showed significant improvement in the overall QOL as well as health after the YBLI, which improvement in the physical, psychological, and social domain scores of WHOQOL-BREF questionnaire. Yoga improved the physical fitness, psychological health, and feeling of general well-being, among the patients practicing in the yoga.

The limitations of the study included lack of an active control group as the non-yoga group was not exposed to any equal attention control intervention and was only on drug therapy as compared to the active yoga intervention group. Hence, the inclusion of such a group would further rule out the therapeutic findings explicitly attributed to the yoga intervention. The number of males was less in both yoga and non-yoga group which was attributed to the higher prevalence of RA in females than males (3:1) (van Vollenhoven, 2009). The change in the levels of psycho-neuro-immune markers was more significant for the women in yoga group as compared to non-yoga group which can also be attributed to small sample size of males in both the groups. The sample size in our study was small; hence the correlation statistics might improve with the inclusion of greater numbers. The study design was single-blinded, which had limitations of an increased likelihood of bias, participant dissatisfaction with non-treatment status, and preconceived notions about treatment. Each session of YBLI lasted for 2 hour every day under the supervision of a certified yoga instructor, which is difficult to sustain regularly for a long term RA management or home practice regimen. For long term sustainable improvements of this chronic debilitating disease, there is need for daily yoga practice and a follow-up period. Future research should explore factors contributing to discontinued practice and whether yoga at home is a suitable substitute for the completion of a yoga program. Further we also plan to elucidate the importance of anti-inflammatory effect of the vagus nerve by measuring the HRV in RA patients after YBLI. In our study, there was no follow-up period, so it was difficult to predict how quickly participants returned to baseline levels of all biomarkers. In future studies, we plan to include a substantial sample size with long term follow-ups to study the long term benefits of the yoga practice.

In conclusion, yoga possesses an immune-modulatory potential which regulates the psycho-neuro-immune axis, reduces disease activity, and improves QOL in RA patients. YBLI maintains immunological tolerance by modulation of inflammatory processes. YBLI also normalizes the dysregulated transcripts associated with the stress pathway. YBLI may be beneficial for RA patients as adjunctive therapy, which enhances physical functioning and improves psychological health.

DATA AVAILABILITY STATEMENT

The datasets generated for this study are available on request to the corresponding author.

ETHICS STATEMENT

The studies involving human participants were reviewed and approved by Institute Ethics Committee, AIIMS, New Delhi, India. The patients/participants provided their written informed consent to participate in this study.

AUTHOR CONTRIBUTIONS

SG designed the study, performed the experiments, analyzed the data, and wrote the manuscript. MK helped SG in standardizing amplification protocols. UK provided the patient samples and clinical data. RD conceived, conceptualized the study, thoroughly reviewed, edited, and finalized the manuscript. All authors contributed to the article and approved the submitted version.

FUNDING

The study was carried out by funds from the Department of Science and Technology (DST), India [SR/SATYAM/55/2016].

CONFLICT OF INTEREST

The authors declare that the research was conducted in the absence of any commercial or financial relationships that could be construed as a potential conflict of interest.

ACKNOWLEDGMENTS

The Senior Research Fellowship (SRF) [45/2/2019-ANA/BMS] to SG by the Indian Council of Medical Research (ICMR), New Delhi, India is thankfully acknowledged. The efforts of Ms. Richa Mishra, yoga instructor who guided and trained the patients on various aspects of yoga are highly appreciated.

REFERENCES

1. AlKhathami, A. D., Alamin, M. A., Alqahtani, A. M., Alsaeed, W. Y., AlKhathami, M. A., and Al-Dhafeeri, A. H. (2017). Depression and anxiety among hypertensive and diabetic primary health care patients. *Saudi Med. J.* 38, 621–628. doi: 10.15537/smj.2017.6.17941
2. Arora, S., and Bhattacharjee, J. (2008). Modulation of immune responses in stress by Yoga. *Int. J. Yoga* 1, 45–55. doi: 10.4103/0973-6131.43541
3. Bauer, R. N., Diaz-Sanchez, D., and Jaspers, I. (2012). Effects of air pollutants on innate immunity: the role of Toll-like receptors and nucleotide-binding oligomerization domain–like receptors. *J. Allergy Clin. Immunol.* 129, 14–26. doi: 10.1016/j.jaci.2011.11.004
4. Bhasin, M. K., Dusek, J. A., Chang, B.-H., Joseph, M. G., Denninger, J. W., Fricchione, G. L., et al. (2013). Relaxation response induces temporal transcriptome changes in energy metabolism, insulin secretion and inflammatory pathways. *PLoS One* 8:e62817. doi: 10.1371/journal.pone.0062817
5. Bonaz, B., Sinniger, V., and Pellissier, S. (2017). The vagus nerve in the neuro-immune axis: implications in the pathology of the gastrointestinal tract. *Front. Immunol.* 8:1452. doi: 10.3389/fimmu.2017.01452
6. Breit, S., Kupferberg, A., Rogler, G., and Hasler, G. (2018). Vagus nerve as modulator of the brain–gut axis in psychiatric and inflammatory disorders. *Front. Psychiatry* 9:44. doi: 10.3389/fpsyt.2018.00044
7. Cimmino, M. A., Salvarani, C., Macchioni, P., Montecucco, C., Fossaluzza, V., Mascia, M. T., et al. (2000). Extra-articular manifestations in 587 Italian patients with rheumatoid arthritis. *Rheumatol. Int.* 19, 213–217. doi: 10.1007/pl00006853
8. Cooper, T. M., McKinley, P. S., Seeman, T. E., Choo, T.-H., Lee, S., and Sloan, R. P. (2015). Heart rate variability predicts levels of inflammatory markers: evidence for the vagal anti-inflammatory pathway. *Brain Behav. Immun.* 49, 94–100. doi: 10.1016/j.bbi.2014.12.017
9. Cutolo, M. (2019). Circadian rhythms and rheumatoid arthritis. *Joint Bone Spine* 86, 327–333. doi: 10.1016/j.jbspin.2018.09.003
10. Cutolo, M., Foppiani, L., and Minuto, F. (2002). Hypothalamic-pituitary-adrenal axis impairment in the pathogenesis of rheumatoid arthritis and polymyalgia rheumatica. *J. Endocrinol. Invest.* 25 10, (Suppl. 1), 19–23. doi: 10.1007/3-540-26861-8_3

11. Cutolo, M., Seriolo, B., Craviotto, C., Pizzorni, C., and Sulli, A. (2003). Circadian rhythms in RA. *Ann. Rheumat. Dis.* 62, 593–596.
12. Cutolo, M., Sulli, A., Paolino, S., and Pizzorni, C. (2016). CTLA-4 blockade in the treatment of rheumatoid arthritis: an update. *Expert Rev. Clin. Immunol.* 12, 417–425. doi: 10.1586/1744666x.2016.1133295
13. Dada, T., Mittal, D., Mohanty, K., Faiq, M. A., Bhat, M. A., Yadav, R. K., et al. (2018). Mindfulness meditation reduces intraocular pressure, lowers stress biomarkers and modulates gene expression in glaucoma: a randomized controlled trial. *J. Glaucoma* 27, 1061–1067. doi: 10.1097/ijg.0000000000001088
14. Duarte-Guterman, P., Yagi, S., Chow, C., and Galea, L. A. M. (2015). Hippocampal learning, memory, and neurogenesis: effects of sex and estrogens across the lifespan in adults. *Horm. Behav.* 74, 37–52. doi: 10.1016/j.yhbeh.2015.05.024
15. Evans, S., Cousins, L., Tsao, J. C., Subramanian, S., Sternlieb, B., and Zeltzer, L. K. (2011). A randomized controlled trial examining Iyengar yoga for young adults with rheumatoid arthritis: a study protocol. *Trials* 12:19.
16. Evans, S., Moieni, M., Lung, K., Tsao, J., Sternlieb, B., Taylor, M., et al. (2013). Impact of Iyengar yoga on quality of life in young women with rheumatoid arthritis. *Clin. J. Pain* 29, 988–997. doi: 10.1097/ajp.0b013e31827da381
17. Firestein, G., and McInnes, I. B. (2017). Immunopathogenesis of rheumatoid arthritis. *Immunity* 46, 183–196. doi: 10.1016/j.immuni.2017.02.006
18. Gagrani, M., Faiq, M. A., Sidhu, T., Dada, R., Yadav, R. K., Sihota, R., et al. (2018). Meditation enhances brain oxygenation, upregulates BDNF and improves quality of life in patients with primary open angle glaucoma: a randomized controlled trial. *Restor. Neurol. Neurosci.* 36, 741–753. doi: 10.3233/rnn-180857
19. Gautam, S., Tolahunase, M., Kumar, U., and Dada, R. (2019). Impact of yoga based mind-body intervention on systemic inflammatory markers and co-morbid depression in active Rheumatoid arthritis patients: a randomized controlled trial. *Restor. Neurol. Neurosci.* 37, 41–59. doi: 10.3233/rnn-180875
20. Gholami, A., Jahromi, L. M., Zarei, E., and Dehghan, A. (2013). Application of WHOQOL-BREF in measuring quality of life in health-care staff. *Int. J. Prev. Med.* 4, 809–817.

21. Glocker, M. O., Guthke, R., Kekow, J., and Thiesen, H.-J. (2006). Rheumatoid arthritis, a complex multifactorial disease: on the way toward individualized medicine. *Med. Res. Rev.* 26, 63–87. doi: 10.1002/med.20045
22. González-Díaz, S. N., Arias-Cruz, A., Elizondo-Villarreal, B., and Monge-Ortega, O. P. (2017). Psychoneuroimmunoendocrinology: clinical implications. *World Allergy Organ J.* 10:19. doi: 10.1186/s40413-017-0151-6
23. Gonzalo-Gil, E., and Galindo-Izquierdo, M. (2014). Role of transforming growth factor-beta (TGF) beta in the physiopathology of rheumatoid arthritis. *Reumatol. Clin.* 10, 174–179. doi: 10.1016/j.reumae.2014.01.006
24. Guo, Q., Wang, Y., Xu, D., Nossent, J., Pavlos, N. J., and Xu, J. (2018). Rheumatoid arthritis: pathological mechanisms and modern pharmacologic therapies. *Bone Res.* 6:15.
25. Hagins, M., States, R., Selfe, T., and Innes, K. (2013). Effectiveness of yoga for hypertension: systematic review and meta-analysis. *Evid. Based Complement. Alternat. Med.* 2013:649836.
26. Hassett, A. L., and Clauw, D. J. (2010). The role of stress in rheumatic diseases. *Arthritis Res. Ther.* 12:123.
27. Jessop, D. S., and Harbuz, M. S. (2005). A defect in cortisol production in rheumatoid arthritis: why are we still looking? *Rheumatology* 44, 1097–1100. doi: 10.1093/rheumatology/keh644
28. Juraska, J. M., Sisk, C. L., and DonCarlos, L. L. (2013). Sexual differentiation of the adolescent rodent brain: hormonal influences and developmental mechanisms. *Horm. Behav.* 64, 203–210. doi: 10.1016/j.yhbeh.2013.05.010
29. Koopman, F. A., Chavan, S. S., Miljko, S., Grazio, S., Sokolovic, S., Schuurman, P. R., et al. (2016). Vagus nerve stimulation inhibits cytokine production and attenuates disease severity in rheumatoid arthritis. *Proc. Natl. Acad. Sci. U.S.A.* 113, 8284–8289. doi: 10.1073/pnas.1605635113
30. Koopman, F. A., van Maanen, M. A., Vervoordeldonk, M. J., and Tak, P. P. (2017). Balancing the autonomic nervous system to reduce inflammation in rheumatoid arthritis. *J. Intern. Med.* 282, 64–75. doi: 10.1111/joim.12626

31. Lawrence, T. (2009). The nuclear factor NF-kappaB pathway in inflammation. *Cold Spring Harb. Perspect. Biol.* 1:a001651.

32. Li, A. W., and Goldsmith, C.-A. W. (2012). The effects of yoga on anxiety and stress. *Altern. Med. Rev.* 17, 21–35.

33. Malm, K., Bergman, S., Andersson, M. L., Bremander, A., and Larsson, I. (2017). Quality of life in patients with established rheumatoid arthritis: a phenomenographic study. *SAGE Open Med.* 5:2050312117713647.

34. Malmström, V., Catrina, A. I., and Klareskog, L. (2017). The immunopathogenesis of seropositive rheumatoid arthritis: from triggering to targeting. *Nat. Rev. Immunol.* 17, 60–75. doi: 10.1038/nri.2016.124

35. Maninger, N., Wolkowitz, O. M., Reus, V. I., Epel, E. S., and Mellon, S. H. (2009). Neurobiological and neuropsychiatric effects of dehydroepiandrosterone (DHEA) and DHEA sulfate (DHEAS). *Front. Neuroendocrinol.* 30:65–91. doi: 10.1016/j.yfrne.2008.11.002

36. Margaretten, M., Barton, J., Julian, L., Katz, P., Trupin, L., Tonner, C., et al. (2011). Socioeconomic determinants of disability and depression in patients with rheumatoid arthritis. *Arthritis Care Res.* 63, 240–246. doi: 10.1002/acr.20345

37. Martinez, G. J., Nurieva, R. I., Yang, X. O., and Dong, C. (2008). Regulation and function of proinflammatory TH17 cells. *Ann. N. Y. Acad. Sci.* 1143, 188–211. doi: 10.1196/annals.1443.021

38. Matcham, F., Rayner, L., Steer, S., and Hotopf, M. (2013). The prevalence of depression in rheumatoid arthritis: a systematic review and meta-analysis. *Rheumatology* 52, 2136–2148. doi: 10.1093/rheumatology/ket169

39. Middleton, K. R., Ward, M. M., Haaz, S., Velummylum, S., Fike, A., Acevedo, A. T., et al. (2013). A pilot study of yoga as self-care for arthritis in minority communities. *Health Qual. Life Outcomes* 11:55. doi: 10.1186/1477-7525-11-55

40. Naveen, G. H., Varambally, S., Thirthalli, J., Rao, M., Christopher, R., and Gangadhar, B. N. (2016). Serum cortisol and BDNF in patients with major depression-effect of yoga. *Int. Rev. Psychiatry* 28, 273–278. doi: 10.1080/09540261.2016.1175419

41. Nishimoto, N., and Takagi, N. (2010). Assessment of the validity of the 28-joint disease activity score using erythrocyte sedimentation rate (DAS28-ESR) as a disease activity index of rheumatoid arthritis in the

efficacy evaluation of 24-week treatment with tocilizumab: subanalysis of the SATORI study. *Mod. Rheumatol.* 20, 539–547. doi: 10.3109/s10165-010-0328-0

42. Otmishi, P., Gordon, J., El-Oshar, S., Li, H., Guardiola, J., Saad, M., et al. (2008). Neuroimmune interaction in inflammatory diseases. *Clin. Med. Circ. Respirat. Pulm. Med.* 2, 35–44.
43. Pascoe, M. C., Thompson, D. R., and Ski, C. F. (2017). Yoga, mindfulness-based stress reduction and stress-related physiological measures: a meta-analysis. *Psychoneuroendocrinology* 86, 152–168. doi: 10.1016/j.psyneuen.2017.08.008
44. Peterson, R. L., Tran, M., Koffel, J., and Stovitz, S. D. (2017). Statistical testing of baseline differences in sports medicine RCTs: a systematic evaluation. *BMJ Open Sport Exerc. Med.* 3:e000228. doi: 10.1136/bmjsem-2017-000228
45. Polsgrove, M. J., Eggleston, B. M., and Lockyer, R. J. (2016). Impact of 10-weeks of yoga practice on flexibility and balance of college athletes. *Int. J. Yoga* 9, 27–34. doi: 10.4103/0973-6131.171710
46. Rajbhoj, P. H., Shete, S. U., Verma, A., and Bhogal, R. S. (2015). Effect of yoga module on pro-inflammatory and anti-inflammatory cytokines in industrial workers of lonavla: a randomized controlled trial. *J. Clin. Diagn. Res.* 9, CC01–CC05.
47. Raveendran, A. V., Deshpandae, A., and Joshi, S. R. (2018). Therapeutic role of yoga in type 2 diabetes. *Endocrinol. Metab.* 33, 307–317. doi: 10.3803/enm.2018.33.3.307
48. Rima, D., Shiv, B. K., Bhavna, C., Shilpa, B., and Saima, K. (2016). Oxidative stress induced damage to paternal genome and impact of meditation and yoga - can it reduce incidence of childhood cancer? *Asian Pac. J. Cancer Prev.* 17, 4517–4525.
49. Sanjabi, S., Zenewicz, L. A., Kamanaka, M., and Flavell, R. A. (2009). Anti- and pro-inflammatory roles of TGF-β, IL-10, and IL-22 in immunity and autoimmunity. *Curr. Opin. Pharmacol.* 9, 447–453. doi: 10.1016/j.coph.2009.04.008
50. Sankalp, Dada, T., Yadav, R. K., and Faiq, M. A. (2018). Effect of yoga-based ocular exercises in lowering of intraocular pressure in glaucoma patients: an affirmative proposition. *Int. J. Yoga* 11, 239–241. doi: 10.4103/ijoy.ijoy_55_17

51. Schmidt, P. J., Daly, R. C., Bloch, M., Smith, M. J., Danaceau, M. A., St Clair, L. S., et al. (2005). Dehydroepiandrosterone monotherapy in midlife-onset major and minor depression. *Arch. Gen. Psychiatry* 62, 154–162. doi: 10.1001/archpsyc.62.2.154

52. Schwarz, L., and Kindermann, W. (1992). Changes in beta-endorphin levels in response to aerobic and anaerobic exercise. *Sports Med.* 13, 25–36. doi: 10.2165/00007256-199213010-00003

53. Shete, S. U., Verma, A., Kulkarni, D. D., and Bhogal, R. S. (2017). Effect of yoga training on inflammatory cytokines and C-reactive protein in employees of small-scale industries. *J. Educ. Health Promot.* 6:76.

54. Sierakowski, S., and Cutolo, M. (2011). Morning symptoms in rheumatoid arthritis: a defining characteristic and marker of active disease. *Scand. J. Rheumatol. Suppl.* 125, 1–5. doi: 10.3109/03009742.2011.566433

55. Silverman, M. N., Pearce, B. D., Biron, C. A., and Miller, A. H. (2005). Immune modulation of the hypothalamic-pituitary-adrenal (HPA) axis during viral infection. *Viral Immunol.* 18, 41–78. doi: 10.1089/vim.2005.18.41

56. Su, H., Lei, C.-T., and Zhang, C. (2017). Interleukin-6 signaling pathway and its role in kidney disease: an update. *Front. Immunol.* 8:405. doi: 10.3389/fimmu.2017.00405

57. Telles, S., and Singh, N. (2012). Is yoga a suitable treatment for rheumatoid arthritis: current opinion. *Open Access J. Sports Med.* 3, 81–87. doi: 10.2147/oajsm.s25707

58. Tolahunase, M. R., Sagar, R., Faiq, M., and Dada, R. (2018). Yoga- and meditation-based lifestyle intervention increases neuroplasticity and reduces severity of major depressive disorder: a randomized controlled trial. *Restor. Neurol. Neurosci.* 36, 423–442. doi: 10.3233/rnn-170810

59. Tolahunase, M., Sagar, R., and Dada, R. (2017). Impact of yoga and meditation on cellular aging in apparently healthy individuals: a prospective, open-label single-arm exploratory study. *Oxid. Med. Cell. Longev.* 2017: 7928981.

60. van Vilsteren, M., Boot, C. R. L., Knol, D. L., van Schaardenburg, D., Voskuyl, A. E., Steenbeek, R., et al. (2015). Productivity at work and quality of life in patients with rheumatoid arthritis. *BMC Musculoskelet. Disord.* 16:107. doi: 10.1186/s12891-015-0562-x

61. van Vollenhoven, R. F. (2009). Sex differences in rheumatoid arthritis: more than meets the eye. *BMC Med.* 7:12. doi: 10.1186/1741-7015-7-12

62. Venihaki, M., Dikkes, P., Carrigan, A., and Karalis, K. P. (2001). Corticotropin-releasing hormone regulates IL-6 expression during inflammation. *J. Clin. Invest.* 108, 1159–1166. doi: 10.1172/jci12869
63. Vijayaraghava, A., Doreswamy, V., Narasipur, O. S., Kunnavil, R., and Srinivasamurthy, N. (2015). Effect of yoga practice on levels of inflammatory markers after moderate and strenuous exercise. *J. Clin. Diagn. Res.* 9, CC08–CC12.

Chapter 6

Online Yoga to Reduce Post Traumatic Stress in Women Who Have Experienced Stillbirth: A Randomized Control Feasibility Trial

Jennifer Huberty[1], Mariah Sullivan[1], Jeni Green[1], Jonathan Kurka[1], Jenn Leiferman[2], Katherine Gold[3] and Joanne Cacciatore[1]

[1] Arizona State University, Tempe, USA.

[2] Colorado School of Public Health, Denver, USA.

[3] University of Michigan Medical School, Ann Arbor, USA.

Citation: Online yoga to reduce post traumatic stress in women who have experienced stillbirth: a randomized control feasibility trial Jennifer Huberty, Mariah Sullivan, Jeni Green, Jonathan Kurka, Jenn Leiferman, Katherine Gold and Joanne Cacciatore. Huberty et al. BMC Complementary Medicine and Therapies (2020) 20:173 https://doi.org/10.1186/s12906-020-02926-3.

ABSTRACT

Background

About 1 in every 150 pregnancies end in stillbirth. Consequences include symptoms of post traumatic stress disorder (PTSD), depression, and anxiety. Yoga has been used to treat PTSD in other populations and may improve health outcomes for stillbirth mothers. The purpose of this study was to determine: (a) feasibility of a 12-week home-based, online yoga intervention with varying doses; (b) acceptability of a "stretch and tone" control group; and (c) preliminary efficacy of the intervention on reducing symptoms of PTSD, anxiety, depression, perinatal grief, self-compassion, emotional regulation, mindfulness, sleep quality, and subjective health.

Methods

Participants ($N = 90$) were recruited nationally and randomized into one of three groups for yoga or exercise (low dose (LD), 60 min per week; moderate dose (MD), 150 min per week; and stretch-and-tone control group (STC)). Baseline and post-intervention surveys measured main outcomes (listed above). Frequency analyses were used to determine feasibility. Repeated measures ANCOVA were used to determine preliminary efficacy. Multiple regression analyses were used to determine a dose-response relationship between minutes of yoga and each outcome variable.

Results

Over half of participants completed the intervention ($n = 48/90$). Benchmarks ($\geq$70% reported >75% satisfaction) were met in each group for satisfaction and enjoyment. Participants meeting benchmarks (completing >90% of prescribed minutes 9/12 weeks) for LD and MD groups were 44% ($n = 8/18$) and 6% ($n = 1/16$), respectively. LD and MD groups averaged 44.0 and 77.3 min per week of yoga, respectively. The MD group reported that 150 prescribed minutes per week of yoga was too much. There were significant decreases in PTSD and depression, and improvements in self-rated health at post-intervention for both intervention groups. There was a significant difference in depression scores ($p = .036$) and grief intensity ($p = .009$) between the MD and STC groups. PTSD showed non-significant decreases of 43% and 56% at post-intervention in LD and MD groups, respectively (22% decrease in control).

Conclusions

This was the first study to determine the feasibility and preliminary efficacy of an online yoga intervention for women after stillbirth. Future research warrants a randomized controlled trial.

Trial registration

ClinicalTrials.gov. NCT02925481. Registered 10–04-16.

BACKGROUND

Every year 2.6 million babies are stillborn worldwide, and one in 150 babies die before or during birth in the United States (U.S.) alone [1]. Stillbirth is defined in the U.S. as the death of a baby from 20-weeks gestation to birth [2, 3]. Stillbirth is unique because it involves birth and death in a single, often very traumatic, event. Bereaved mothers are seven times more likely to to have symptoms of post-traumatic stress disorder (PTSD), four times more likely to be diagnosed with clinical depression, and twice as likely to be diagnosed with an anxiety disorder [4]. Symptoms of PTSD have been reported to last anywhere from 2 months to 18 years [5]. The long-lasting adverse effects of stillbirth may even impact the health of the mother and child in subsequent pregnancies.

Women often become pregnant relatively quickly after a stillbirth (50–98% of pregnancies within 12–18 months) and yet may not be emotionally or physically ready for another child [6, 7]. This may lead to dysregulation of family life leading to psychological and other problems in subsequently born children [8]. Mothers who have experienced stillbirth are more likely to report symptoms of anxiety (22.5%) and depression (19.7%) during subsequent pregnancies compared to women whose babies are born alive [9]. In addition, poor and emotional health outcomes have been associated with adverse health outcomes (i.e., poor self-rated health) which can affect physical well-being in subsequent pregnancies [5]. Thus, there is a need for inter-conception care (time between stillbirth and the subsequent pregnancy or decision not to conceive again) for bereaved mothers after the stillbirth of a baby.

There is currently no standard of care during the inter-conception period after stillbirth. A recent systematic review of experimental interventions revealed only two interventions to improve emotional, mental, or physical health after a stillbirth in the last 36 years [10]. In one study, families met with a grief support team twice within 6 months and there were no differences in grief at post intervention compared to the control [11]. The second study was a mindfulness-based feasibility study for anxiety and depression with two sessions over a two-week time period but only six participants completed both sessions, limiting the generalizability of the findings [12]. Mothers often report seeking help through support groups, psychotherapy, and medication [13] to help cope with the stressors of traumatic grief. Considering the negative health implications following stillbirth and risks of poor mental, emotional, or physical health in subsequent pregnancies, interventions are desperately needed.

Yoga may be a feasible intervention to improve women's well-being during the inter-conception period after stillbirth. Yoga is a non-phamacologic option for women who want to avoid medication in the inter-conception period due to concern for the subsequent baby and possible adverse effects. Yoga brings together the practice of breathing, physical postures, and meditation [14] and has been utilized for decades to reduce stress and disease [15]. It has been reported to reduce symptoms of PTSD, anxiety, and depressive symptoms in a number of populations [15, 16], and it is a complementary approach to reduce stress during pregnancy and aiding in postpartum depression [17]. In a small non-randomized trial in mothers after stillbirth, Huberty and colleagues reported the majority of participants were satisfied with a 60 min per week online yoga intervention with reported benefits such as reducing mental and physical stress, improved mindfulness and calm, and getting more time for themselves [18].

While the exact mechanism through which yoga reduces symptoms of PTSD, anxiety, and depression is still unknown, changes may be attributable, at least in part, to improvements in mindfulness, emotional regulation, self-compassion, and/or sleep quality. For example, bereaved mothers often report recurring flashbacks, hypervigilance, hyperarousal, rumination, and avoidance [19]. Through the cultivation of mindfulness, yoga provides the practitioner with a means to focus attention and increase awareness of the mind and body. Enhanced mindfulness may increase acceptance of one's own emotions, improving regulation and reducing avoidance [20]. Yoga may also bring a deeper awareness of negative beliefs about oneself (such as self-criticism) [20] and promote self-compassion [21,22,23,24].

Yoga may also improve sleep quality. A reduction in ruminating thoughts, hypervigilance, and hyperarousal may lower cortisol levels in the body, promoting parasympathetic nervous system activity [22, 25, 26]. This improves both the ability to fall asleep and stay asleep. This is important as studies suggest dysregulated sleep patterns may predict PTSD symptoms [27]. Yoga may improve comorbid conditions associated with PTSD including depression and anxiety, giving it high utility as an adjunctive treatment to cope with grief after stillbirth, especially during the interconception period [28].

Yoga has the potential to improve mental, emotional, and physical health in these bereaved mothers, hence, the purpose of this study was to determine the feasibility (i.e., acceptability, demand) of a 12-week, home-based, online-streamed yoga intervention with varying doses (low = 60 min per week; moderate = 150 min per week). We also explored the acceptability of an active control group to be used in a future efficacy trial. Finally, we determined the preliminary effects (not powered for efficacy) of the intervention on reducing symptoms of PTSD, anxiety, depression, and perinatal grief and the proposed mechanism of self-compassion, emotional regulation, mindfulness, sleep quality, and self-rated health.

METHODS

Study Design

This was a three-group randomized feasibility trial. The Institutional Review Board at Arizona State University approved this study. All participants provided their consent to participate and granted the research team permission to collect medical records before beginning the study. This study adheres to CONSORT guidelines.

Participant Recruitment and Selection

Participants were recruited using snowball sampling through stillbirth-related non-profit organizations (e.g., Stories of Babies Born Still, Star Legacy Foundation, Mothers in Sympathy and Support Foundation), social media (e.g., Facebook, Instagram), and hospital/clinics. Research staff asked organizations to advertise the study by posting recruitment information (e.g., fliers, memes) in clinics or to their social media sites, websites, newsletters, and listservs. Participants could receive up to $65 for completing each component of the intervention (e.g., questionnaires, interviews). Recruitment occurred from July 2017 until September 2018, and the intervention occurred from August 2017 until May 2018. The intervention concluded when all participants ($N = 90$, 30 per group) had completed the intervention.

As this was a feasibility study, the sample was appropriate for early phase-development trials and consistent with previously conducted beta tests that demonstrated a 20% attrition rate [29]. Relatively small sample sizes are common in feasibility studies [29], thus a post-hoc sample size analysis was conducted based on observed effect size of the primary outcome PTSD.

Interested participants were directed to a web-based eligibility screener via Qualtrics. See Table 1 for eligibility criteria. Though pregnant women were excluded from the study, if a woman became pregnant during the study she could still participate if she completed a PARmed-X for Pregnancy (a screening tool for prenatal exercise) and received approval from her physician. Ineligible participants were notified by phone or email and were offered a discounted membership to Udaya, the yoga platform used for this study. Eligible participants were sent an email to schedule an intake phone call to explain study procedures, the informed consent, and the HIPAA authorization form. After the intake phone call, participants were sent a link to sign an informed consent document and complete a baseline questionnaire. The research team faxed a copy of the signed informed consent and HIPAA authorization form to the physician's office to release the participants' medical records to the research team.

Table 1. Inclusion and exclusion criteria

Inclusion	**Exclusion**
Experienced a stillbirth within the past 6 weeks to 24 months	Unstable psychiatric condition (i.e., psychosis; suicidal ideation with plan)
Clinical levels of posttraumatic stress symptoms (score of ≥33 on the Impact of Events Scale)	Pregnant at time of enrollment
≥18 years of age	Practiced yoga at least 60 min/week
Resided in the U.S.	Scored 20–27 on the Patient Health Questionnaire-9 (i.e., severe depression)
Able to read/understand/speak English	At risk for suicide based on follow-up phone assessment after positive screen (score of 1, 2, or 3 on the last question on the PHQ-9)
Underactive (≤120 min/week of moderate intensity physical activity)	
Willing to be randomized	

Regular internet access via mobile phone, desktop/laptop computer, tablet etc.	
Answered "no" to all items on the PAR-Q (i.e., can participate in exercise safely) or given clearance by a doctor	

Randomization

Following baseline assessments, women were enrolled into one of the following three groups for 12 weeks: 1) intervention low dose (LD) = 60 min/week yoga (n =30), 2) intervention moderate dose (MD)= 150 min/week yoga (n =30), or 3) stretch and tone control group (STC)= 60 min/week of stretching/toning exercises (n =30). A list of randomized participant numbers were generated using randomizer.org by a colleague who was not a member of the research team. The list was concealed from members of the research team and revealed only when it had been assigned to a participant. Members of the research team would then enroll participants based on their assigned numbers.

Procedures

Once participants were assigned to a group, the research team mailed the intervention group a package containing study information and directions, one yoga mat, two blocks (i.e., brick-shaped prop to assist in reaching the floor), and one yoga strap (i.e., long cloth to help increase range of motion). Participants in the control group were also sent a package with study information and directions, a stretching mat, and one resistance band. Participants were asked to notify the research team once they received their package. Once the research team was notified, participants were provided an on-line username and password to begin the intervention via email.

Intervention

The intervention groups (LD and MD) were asked to follow a 12-week online yoga prescription developed by the research team [18]. Briefly, the prescription included 12 videos developed for women who had experienced stillbirth [18] and the remaining 48 videos were chosen using Udaya's existing library of more than 600 yoga videos. Udaya is an online streaming platform that provides yoga videos that vary in length, style, and language, instructed by experienced yoga teachers (e.g., 200 h or more of yoga training).

All videos were appropriate for women up to 20 weeks gestation. Udaya yoga videos were Hatha based, which combines a physical practice with a meditative component to cultivate mindfulness (i.e., moment-to-moment, non-judgmental awareness, cultivated by paying attention in specific ways) [30] and self-awareness. Hatha yoga includes instruction from a variety of lineages (e.g., Iyengar, Ashtanga, Vinyasa). Both the LD and MD groups had identical poses and sequences in each prescription. Participants were asked to complete the yoga videos in a specific order, which included specialized preparatory instructions for each video to ensure comfort and safety.

Control

Participants in the stretch and tone control (STC) group were asked to follow a 12-week online stretching/toning exercise prescription for 60 min per week to match the LD intervention group. The research team developed 12, 30-min videos for the STC group prescription (produced and filmed by Udaya). These videos were developed by adapting a well-established evidence-based protocol specific for women who are underactive [30]. Each video included a three-minute warm up and cool down, and a stretch or tone exercise for 1–2 sets of 20–45 s or 10–15 repetitions.

Measures

All data were collected online via Qualtrics.

Demographics

Demographic data collected at baseline included date of birth, race/ethnicity, household income, education level, marital status, occupation, height and weight. Participants were asked if they had a history of chronic health conditions (e.g., hypertension), history of post-traumatic stress disorder and/or clinical depression, were were currently on medications for mental health, and for a rating of their overall physical health. Finally, participants were asked whether or not they were currently participating in any in-person and/or online support groups.

Feasibility

All feasibility measures were assessed at post-intervention (12-weeks). We measured feasibility using Bowen's guidelines [31] including acceptability, demand, and preliminary efficacy (i.e., *trends* in changes on reducing symptoms of PTSD, anxiety, depression, and perinatal grief and the proposed mechanism of self-compassion, emotional regulation, mindfulness, sleep quality, and self-rated health.

Acceptability

We assessed acceptability by ability to perform/comply with the intervention and satisfaction. Ability to perform/comply was measured using number of dropouts and number of participants who completed each part of the intervention (i.e., daily and weekly logs, self-reported outcomes, and satisfaction survey). Participants were asked to self-report their online yoga/exercise sessions in an online daily log including which video they completed, date, and start/end times. Satisfaction was measured with an investigator-developed survey of 29 questions (e.g., Likert-scale, yes/no, options). Benchmarks for satisfaction related to enjoyment, intervention, and instruction/instructors were that at least 70% of participants would report 75% satisfaction (see Table 2).

Table 2. Satisfaction

Question	**Response**	*n*
Enjoyment		
Overall, how much did you enjoy your online classes?	Low dose group	
	Very much	11 (61)
	Somewhat	6 (33)
	Not at all	1 (6)
	Moderate Dose	
	Very much	7 (50)
	Somewhat	7 (50)
	Not at all	-
	Control	
	Very much	5 (36)
	Somewhat	9 (64)
	Not at all	-

Did you enjoy doing exercise/yoga at home?	Low Dose	
	Yes	17 (94)
	No	1 (6)
	Moderate Dose	
	Yes	14 (100)
	No	-
	Control	
	Yes	13 (93)
	No	1 (7)
Prescription		
Were you able to complete your prescribed amount (i.e., 60 or 150 minutes) of online videos each week?	Low Dose	
	Yes	12 (67)
	No	6 (33)
	Moderate Dose	
	Yes	3 (20)
	No	12 (80)
	Control	
	Yes	7 (50)
	No	7 (50)
I felt the prescribed minutes (i.e., 60 or 150 minutes) per week of online videos was:	Low Dose	
	Too much	1 (6)
	About right	15 (88)
	Too little	1 (6)
	Moderate Dose	
	Too much	8 (57)
	About right	6 (43)
	Too little	-
	Control	
	Too much	-
	About right	10 (71)
	Too little	4 (29)
Satisfaction		

I am satisfied with participating in the online streaming videos	Low Dose	
	Strongly agree	7 (39)
	Agree	8 (44)
	Neutral	2 (11)
	Disagree	1 (6)
	Moderate Dose	
	Strongly agree	8 (57)
	Agree	5 (36)
	Neutral	-
	Disagree	1 (7)
	Control	
	Strongly agree	9 (64)
	Agree	5 (36)
	Neutral	-
	Disagree	-
I am satisfied with the instructor(s) in the online videos	Low Dose	
	Strongly agree	7 (39)
	Agree	6 (33)
	Neutral	4 (22)
	Disagree	1 (6)
	Moderate Dose	
	Strongly agree	9 (64)
	Agree	5 (36)
	Neutral	-
	Disagree	-
	Control	
	Strongly agree	10 (71)
	Agree	4 (29)
	Neutral	-
	Disagree	-
Coping		

The online streaming videos has helped me to cope with the grief associated with the death of my baby	Low Dose	
	Strongly agree	5 (28)
	Agree	7 (39)
	Neutral	4 (22)
	Disagree	2 (11)
	Moderate Dose	
	Strongly agree	5 (36)
	Agree	5 (36)
	Neutral	3 (21)
	Disagree	1 (7)
	Control	
	Strongly agree	3 (21)
	Agree	5 (36)
	Neutral	4 (29)
	Disagree	2 (14)
Easiness		
The classes were easy	Low Dose	
	Strongly agree	4 (19)
	Agree	5 (28)
	Neutral	6 (33)
	Disagree	3 (17)
	Moderate Dose	
	Strongly agree	-
	Agree	6 (43)
	Neutral	6 (43)
	Disagree	2 (14)
	Control	
	Strongly agree	-
	Agree	8 (57)
	Neutral	4 (29)
	Disagree	2 (14)

The instruction in the videos was easy to follow	Low Dose	
	Strongly agree	6 (33)
	Agree	9 (50)
	Neutral	2 (11)
	Disagree	1 (6)
	Moderate Dose	
	Strongly agree	4 (29)
	Agree	9 (64)
	Neutral	1 (7)
	Disagree	-
	Control	
	Strongly agree	11 (79)
	Agree	3 (21)
	Neutral	-
	Disagree	-
It was easy to find the time to do the videos	Low Dose	
	Strongly agree	2 (11)
	Agree	8 (44)
	Neutral	4 (22)
	Disagree	3 (17)
	Strongly disagree	1 (6)
	Moderate Dose	
	Strongly agree	-
	Agree	3 (21)
	Neutral	3 (21)
	Disagree	6 (43)
	Strongly disagree	2 (14)
	Control	
	Strongly agree	1 (7)
	Agree	5 (36)
	Neutral	4 (29)
	Disagree	3 (21)
	Strongly disagree	1 (7)
Continue/Recommend		

I will continue participating in online streaming exercise or yoga in the future	Low Dose	
	Strongly agree	2 (11)
	Agree	11 (61)
	Neutral	4 (22)
	Disagree	-
	Strongly disagree	1 (6)
	Moderate Dose	
	Strongly agree	7 (50)
	Agree	2 (14)
	Neutral	3 (21)
	Disagree	-
	Strongly disagree	2 (14)
	Control	
	Strongly agree	4 (29)
	Agree	7 (50)
	Neutral	2 (14)
	Disagree	1 (7)
	Strongly disagree	-
I will recommend participating in online streaming exercise or yoga to other women who have experienced stillbirth	Low Dose	
	Strongly agree	8 (44)
	Agree	9 (50)
	Neutral	1 (6)
	Disagree	-
	Moderate Dose	
	Strongly agree	9 (64)
	Agree	2 (14)
	Neutral	2 (14)
	Disagree	1 (7)
	Control	
	Strongly agree	8 (57)
	Agree	5 (36)
	Neutral	-
	Disagree	1 (7)
Social Support		

If we added an online social support group, how would it have affected your participation?	Low Dose	
	Would have participated in MORE videos	5 (50)
	It wouldn't have changed anyting	2 (20)
	I don't know	3 (30)
	Would have participated in LESS videos	-
	Moderate Dose	
	Would have participated in MORE videos	6 (86)
	It wouldn't have changed anyting	1 (14)
	I don't know	-
	Would have participated in LESS videos	-
	Control	
	Would have participated in MORE videos	5 (71)
	It wouldn't have changed anyting	-
	I don't know	2 (29)
	Would have participated in LESS videos	-
If we were to conduct a future study, do you think we should add an online social support group (i.e., Facebook group, discussion board)?	Low Dose	
	Yes	11 (100)
	No	-
	Moderate Dose	
	Yes	7 (88)
	No	1 (12)
	Control	
	Yes	7 (100)
	No	-

Demand

Demand was determined by participants reporting they would recommend yoga to others, continue to participate, and that the program was easy to complete. Benchmarks were that at least 70% of participants would report they agree to each construct, measured with the satisfaction survey. Demand was also determined using attendance to the intervention; analytics tracking from the Udaya website via Wistia plugin, a software that tracks engagement with online videos [29]. Data collected included title of the video, date and time it was completed, number of minutes per session, total number of minutes completed each week. Number of yoga activity minutes per week were compiled primarily from self-report logs while supplementing the Wistia plugin values when self-report logs were missing. Benchmarks for the intervention groups were set for participants to complete at least 90% of the respective group's prescribed weekly dose of yoga minutes (LD= 55 min; MD= 135 min) on at least 9 out of the 12 weeks (75%).

Preliminary Efficacy

All preliminary efficacy measures were collected at baseline, post-intervention (12-weeks), and follow-up (20-weeks).

Impact of Event Scale

Post-traumatic stress disorder symptoms were measured using the Impact of Event Scale (IES-R). The IES-R was developed to reflect the criteria for PTSD per the Diagnostic Symptom Manual (DSM-IV-TR). The scale consists of 22 questions which are scored on a five-point Likert scale (0=not at all, 1=a little bit, 2=moderately, 3=quite a bit, 4=extremely). There are three subscales (i.e., avoidance, intrusion, hyperarousal) and the sum of the three subscales scores comprise the total score. Other studies have reported subscale scores to have high internal consistency (α=0.79–0.92). A total score≥33 indicates the "best cutoff for a probable diagnosis for PTSD". Possible scores range from 0 to 88 [32].

State-Trait Anxiety Inventory

The State-Trait Anxiety Inventory (STAI) yields scores indicating levels of trait (Form Y-1) and state (Form Y-2) anxiety, with higher scores indicating greater levels of anxiety. This scale is scored using a four-point Likert scale (1=not at all, 2=somewhat, 3=moderately so, 4=very much so), and participants respond to questions such as, "I feel calm," and "I feel nervous and restless." The STAI has demonstrated reliability in pregnant and postpartum populations. Possible scores range from 20 to 80 [33].

Patient Health Questionnaire-9

The Patient Health Questionnaire-9 (PHQ-9) was used to measure depressive symptoms. This measure is commonly used to screen, diagnose, monitor, and measure the severity of depression. Scores range from 0 to 27 and cut-off scores of 5, 10, 15, and 20 indicate mild, moderate, moderately severe, and severe depressive symptoms, respectively. This scale is valid and reliable in general populations ($\alpha=.86$–$.89$) and has been validated in pregnant and postpartum populations [34, 35].

Perinatal Grief Scale

The Perinatal Grief Scale (PGS) was used to measure symptoms of grief after perinatal loss. It is a valid 33-item instrument with three subscales (i.e., active grief, difficulty coping, and despair) with higher scores indicating more intense perinatal grief. Possible scores range from 33 to 165 [36].

Self-Compassion Scale

The Self-Compassion Scale (SCS) is a 26-item questionnaire using a five-point Likert scale from 1=almost never to 5=almost always, The SCS consists of six subscales (self-kindness, self-judgment, common humanity, isolation, mindfulness, over-identified) and has good construct validity and reliability ($\alpha=.92$). A total score is calculated by taking the mean of each subscale and reverse scoring the negative subscale items (i.e., self-judgment, isolation, over-identification) and computing a total mean. Higher scores indicate higher levels of self-compassion and possible scores range from 1 to 5. A sample question includes, "When times are really difficult, I tend to be tough on myself," [37].

Emotion Regulation Questionnaire

The Emotion Regulation Questionnaire (ERQ) is a 10-item scale used to measure an individual's tendency to regulate his or her emotions by two strategies: cognitive reappraisal and expressive suppression. Most alphas exceed .80. Higher scores indicate greater use of emotional regulation strategies [38, 39].

Mindful Attention Awareness Scale

The Mindful Attention Awareness Scale (MAAS) is a 15-item scale that measures the extent to which individuals are able to maintain awareness of present-moment experience. This scale uses a 6-point Likert scale (ranging from 1=almost always to 6=almost never), and the mean is computed to generate a total score. Higher scores indicate higher levels of mindfulness and possible scores range from 1 to 6. This scale is a valid and reliable measure with good internal consistency (α=.80–.87) [40].

Pittsburgh Sleep Quality Index

Sleep was measured subjectively using the Pittsburgh Sleep Quality Index (PSQI). The PSQI is a 19-item questionnaire that has demonstrated reliability and validity in pregnant populations [41]. Higher global PSQI scores indicate more sleep disturbances and possible scores range from 0 to 21.

SF-12. The SF-12 is a measure of self-reported health developed as a shorter version (12-item) of the original SF-36 Health Survey. The physical health composite score was used in this study. Scores range from 0 to 100, with zero being the lowest level of health and 100 the highest [42].

Data Analysis

Descriptive analyses were performed for demographic characteristics using means and standard deviation of continuous data and frequencies and proportions of discrete data for each intervention group. One-way analysis of variance were conducted to examine between-group differences on demographic characteristics. Frequency analyses were used to determine feasibility (acceptability and demand), compliance, and outcome measures using only participants that completed both baseline and post intervention surveys. To test the preliminary efficacy of the trial, repeated measures (collected at baseline, post-intervention, follow-up time points) analysis of covariance analyses (ANCOVA) were performed for each of the outcome measures of PTSD, anxiety (state and trait), depression, perinatal grief, self-compassion, emotional regulation and suppression, mindfulness attention awareness, self-rated health, and subjective sleep quality (PSQI). Lastly, to investigate the dose of yoga activity related to changes in the outcome measures, multiple regression analyses were conducted using the pooled sample (i.e. no group stratification) for each outcome for all participants with baseline and post-intervention data.

ANCOVA analyses were conducted independently on each outcome while controlling for age, race, household income, education level, marital status, BMI, and the level of peer support received (none, online, in-person, both online and in-person). Regression models were built using a hierarchical approach, adding demographic variables based on theoretical importance while consindering multicollinearity indices. All analyses were conducted while applying appropriate adjustments (i.e. Greenhouse-Gesser corrections when sphericity was not met, Bonferroni corrections for post hoc analyses) to account for non-parametric data and multiple comparisons and effect sizes were computed [43] using SPSS [44] with a statistical significance threshold set at $p < .05$. Post-hoc analysis to determine required sample size to achieve statistical significance ($\alpha < 0.05$) with appropriate power ($1- \beta = 0.80$) for repeated measures within-between interaction effects was conducted using G*Power 3.1.9.4.

RESULTS

Participants

A total of 707 participants completed the eligibility questionnaire. Of those, 196 participants met the eligibility criteria and 94 enrolled in the study. Four enrollees did not start the study, yielding a total of 90 participants. Once eligibile but prior to randomization, 30% of women (n =30) were asked to verify their stillbirth (i.e., doctor's note) with 28 verified and two unverified and ineligible. Sixty participants were randomized to the intervention groups (LD n =30; MD n =30), and 30 participants were randomized to the control group (see Fig. 1).

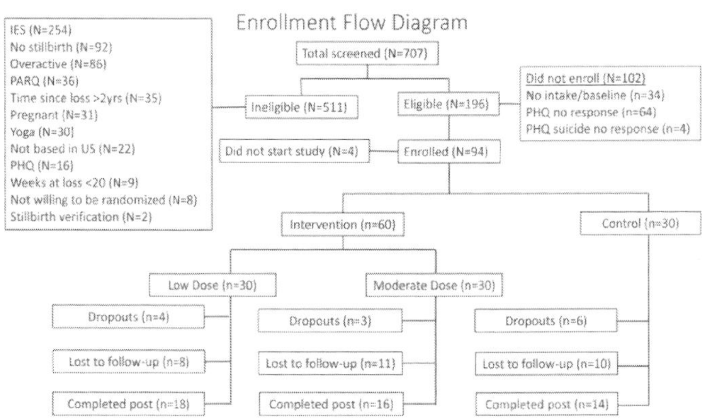

Figure. 1. CONSORT diagram

Demographics

Participants were predominantly White (n =78; 86%). Most participants (n =53; 59%) earned $61,000 per year or more, were college educated (baccalaureate n =32; 36% and graduate n =20; 22%), and married (n =75; 83%). The average gestational age at the time of stillbirth was 30 weeks (SD=6.99). See Table 3 for further demographic information.

Table 3. Participant demographics

	n	%
Race & Ethnicity		
Hispanic	9	10
Non-Hispanic	80	88.9
American Indian or Alaskan Native	1	1.1
Asian or Asian-American	1	1.1
Native Hawaiian or Pacific Islander	1	1.1
White, European-American, Caucasian	78	86.7
Black, African American, Nativ African	5	5.6
Bi-racial or Multi-racial	3	3.3
Other	1	1.1
Income		
Less than $20,000	8	8.9
$21,000-40,000	16	17.8
$41,000-60,000	12	13.3
$61,000 and above	53	58.9
Education		
High school diploma	5	5.6
Some college	20	22.2
Associates	12	13.3
Bachelors degree	32	35.6
Graduate degree or above	20	22.2

Martital Status		
Single	6	6.7
Partnered/In a relationship	5	5.6
Married	75	83.3
Divorced	3	3.3
	Mean	SD
Stillbirth		
Weeks gestiation at time of loss	30.43	6.99
Time since stillbirth	40.92	29.17
	n	%
Delivery method		
Vaginal	79	87.8
Cesarean Section	11	12.2
Know cause of death?		
Yes	47	52.2
No	43	47.8
Did baby survive at all?		
Yes	3	3.3
No	87	96.7

Feasibility

Acceptability

Ability to perform/comply with intervention

Thirteen participants (14%) formally dropped out of the study (LD $n = 4$, MD $n = 3$, STC $n = 5$) and 11 completed a survey about the reasons for withdrawal. Twenty-nine participants did not complete the study (i.e., did not formally drop out, but did not continue to participate in yoga and did not complete the post intervention surveys) (See Fig. 1). Reasons for drop out varied and included mood, pregnancy, time, and stress. Participants who dropped out reported they enjoyed the online streaming videos ($n = 7$;

63.6%) and thought an online social support group would have helped them continue participating in the study (n =7; 55%). More than one-half of participants completed the post-intervention and satisfaction survey (n =48/90; LD n =18; MD n =16; STC n =14) (See Table 4).

Table 4. Ability to comply with intervention components

	Low Dose	Moderate Dose	Control
Daily logs	201/360 (56%)	160/360 (44%)	148/360 (41%)
Weekly logs	201/360 (56%)	160/360 (44%)	148/360 (41%)
Baseline	90/90 (100%)	90/90 (100%)	90/90 (100%)
Post	18/30 (60%)	16/30 (53%)	14/30 (46%)
Dropout	4	3	5
Lost to fol-low up	8	11	10

Satisfaction

Benchmarks (≥70% would report 75% satisfaction) were met for all groups related to satisfaction of the intervention including enjoyment and instruction/instructors. Benchmarks (>70%) for prescribed minutes being appropriate were met in the LD and STC groups but not in the MD group (See Table 2).

Nearly all participants (n =45; 98%) reported that online yoga affected their lives in a positive way due to increased calmness, relaxation, enhanced ability to handle life and process emotions, and more time for self, though reported benefits varied by group (i.e., intervention groups v control). Almost 70% of participants reported they would have participated more if the study included a support group and 96% reported social support should be included in future studies (See Table 2).

Barriers to participation in the study (n =25/48; 52%) included work/ school commitments, care for others, mood, and time. No adverse injuries were reported during this study but there were some reports of discomfort (n =8, 12%), though these included normal symptoms of exercise (e.g., muscle soreness, wrist/ knee discomfort). Seven participants became pregnant during the study (LD n =4, MD n =1, STC n=2) and five of those reported pregnancy was a barrier to participation in the study for reasons including being afraid of hurting the baby and fatigue. Half of all participants who completed the intervention reported experiencing a technical limitation while watching the videos (LD $n = 7$, MD $n = 6$, STC $n = 10$). Across all groups, the most commonly reported technical limitation was bad or slow internet connection ($n = 15$).

Demand

Approximately 60% of participants would recommend online yoga to others who have experienced stillbirth, and that they would continue to participate in online yoga and thought the program was easy to complete. Benchmarks for demand (≥70% in all categories), however, were not met in any group (See Table 2). Benchmarks for attendance (i.e., time spent in yoga; completing at least 90% of the prescribed minutes at least 9/12 weeks) were not met in any of the groups. The percent of participants meeting benchmarks for LD and MD groups were 44% ($n = 8/18$) and 6% (n = 1/16), respectively. Across the intervention, LD and MD groups averaged 44 and 77 min per week of yoga, respectively. Participants over-reported online yoga participation compared to Wistia analytics tracking across the 12-week study.

Preliminary Efficacy

There were no significant differences in demographics or peer-support between groups at the baseline time point. The primary outcome of IES-R had no main effect from intervention group ($F(2,22) = 1.460, p = .254, \eta^2 = 0.117$), but did decrease throughout the trial ($F(1.511, 33.237) = 5.231$, $p = 0.017$, $\eta^2 = 0.192$), specifically from baseline to post-intervention (Fig. 2). There was no significant interaction between IES-R and intervention group, indicating that regardless of intervention group, PTSD was not affected by group assignment. Secondary outcomes of depression (F (2, 21) = 3.90, $p = 0.036$, $\eta^2 = 0.271$) and perinatal grief F (2, 22) = 5.830, $p = 0.009$, $\eta^2 = 0.346$) showed significant effect between intervention groups with expected differences in LD and MD groups compared to the STC. Depression ($F(1.486, 31.20) = 6.788$, $p = 0.007$, $\eta^2 = 0.244$) and self-compassion (F (2, 46) = 3.229, $p = 0.049$, $\eta^2 = 0.123$) showed improvements over the course of the trial (Figs. 3 & 4). Self-rated health ($F(1.580, 34.749) = 3.830$, $p = 0.041$, $\eta^2 = 0.148$) decreased over the course of the trial in the LD and STC groups. Nonsignificant differences between groups indicated MD blunted this self-perceived diminution $F(3.159, 34.749) = 0.592$, $p = 0.633$, $\eta^2 = 0.051$).

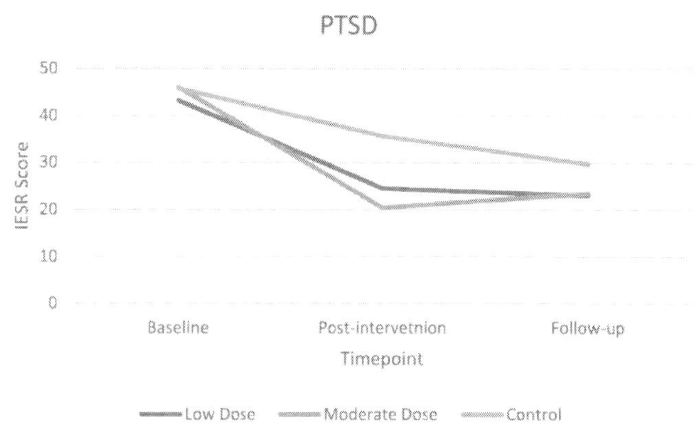

Figure 2. Change in PTSD scores by timepoint.

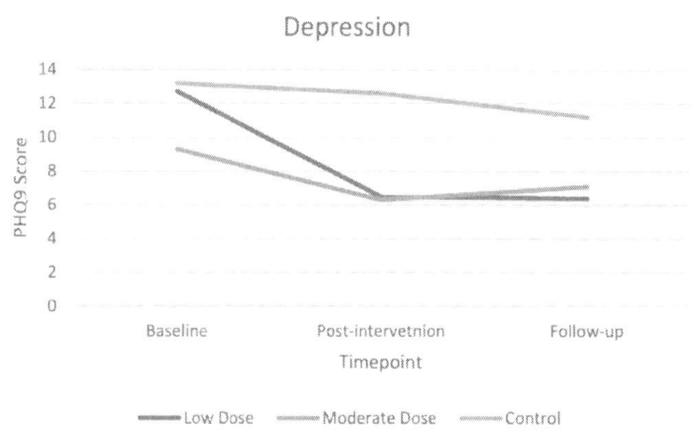

Figure 3. Change in depression by timepoint.

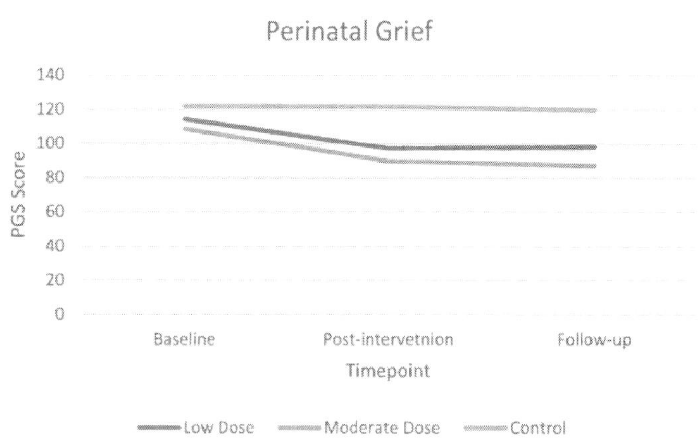

Figure 4. Change in perinatal grief by timepoint.

Although no interactions were found for any outcome between groups over time with effect sizes ranging from 0.01–0.156, the primary outcome of PTSD showed non-significant decreases of 43 and 56% (the largest relative change among outcome measures) at post-intervention for the LD and MD groups, respectively, compared to a 22% decrease in the STC (F (2, 22) = 2.212, p =0.133, partial η^2 =0.167). Post-hoc sample size analysis for the intervention group-by-repeated IES-R measure interaction effect size (η2=0.105) determined that 171 participants (57 per group) would be required to achieve statistical significance. Sleep quality also showed non-significant improvements of 34% in the LD group and 29% in the MD group compared to a 10% worsening in the STC between baseline and post-intervention (F (2, 21) = 3.148, p =0.064, partial η^2 =0.231). Finally, perinatal grief remained static in the STC throughout the trial while MD and LD groups showed non-significant decreases from baseline to post-intervention (F (2, 22) = 1.769, p =0.194, η^2 =0.139) that was maintained at follow-up. Outcomes of anxiety (STAI), self-compassion (SCS), emotional regulation (ERQ), and mindfulness (MAAS) showed visual improvements in the hypothesized direction but no significant differences. See Table 5 for mean differences by timepoint.

Table 5. Mean differences of intervention x timepoint

		Means (95% CI)		
Outcome	**Time point**	**Low Dose**	**Moderate Dose**	**Study Control**
Impact of Events (PTSD)				
	pre-intervention	43.16 (35.77–50.55)	45.99 (39.30–52.68)	45.75 (36.81–54.69)
	post-intervention	24.53 (16.72–32.34)	20.38 (13.31–27.45)	35.67 (26.22–45.12)
	follow-up	23.03 (13.86–32.20)	23.39 (15.08–31.69)	29.67 (18.57–40.77)
Anxiety (STAI)				
	pre-intervention	91.38 (82.78–99.98)	86.68 (79.36–94.01)	90.67 (80.80–100.55)
	post-intervention	87.80 (80.72–94.89)	86.06 (80.03–92.09)	88.77 (80.63–96.90)
	follow-up	88.17 (81.60–94.75)	82.36 (76.76–87.95)	85.19 (77.65–92.74)
Depression (PHQ)				
	pre-intervention	12.67 (9.23–16.11)	9.27 (6.28–12.26)	13.15 (9.14–17.17)

	post-intervention	6.46 (3.50–9.42)	6.31 (3.74–8.89)	12.56 (9.10–16.02)
	follow-up	6.38 (3.57–9.19)	7.11 (4.66–9.56)	11.21 (7.92–14.49)
Self-compassion (SCS)				
	pre-intervention	2.52 (2.06–2.98)	2.79 (2.31–3.28)	2.46 (1.85–3.07)
	post-intervention	2.88 (2.40–3.37)	2.86 (2.35–3.37)	2.61 (1.97–3.26)
	follow-up	2.83 (2.34–3.32)	2.85 (2.33–3.37)	2.54 (1.89–3.20)
Emotional Regulation (ERQ)				
	pre-intervention	17.63 (14.42–20.84)	19.42 (16.38–22.46)	22.62 (18.73–26.51)
	post-intervention	17.75 (14.37–21.12)	16.89 (13.69–20.09)	20.20 (16.10–24.29)
	follow-up	16.10 (11.91–20.30)	19.06 (15.09–23.03)	18.29 (13.20–23.37)
Self-rated Health (SF12)				
	pre-intervention	61.19 (56.79–65.60)	57.03 (53.04–61.02)	59.05 (53.72–64.38)
	post-intervention	57.26 (52.94–61.57)	57.24 (53.33–61.15)	58.04 (52.81–63.26)
	follow-up	56.08 (51.32–60.84)	57.14 (52.83–61.45)	55.93 (50.17–61.69)
Perinatal Grief (PGS)				
	pre-intervention	114.42 (103.51–125.34)	108.70 (98.82–118.59)	122.11 (108.91–135.31)
	post-intervention	97.21 (84.31–110.12)	89.70 (78.01–101.38)	121.77 (106.15–137.38)
	follow-up	98.12 (81.05–115.19)	86.88 (71.42–102.35)	119.73 (99.07–140.39)
Sleep Quality (PSQI)				
	pre-intervention	11.40 (9.13–13.67)	8.87 (6.72–11.01)	8.26 (5.51–11.01)
	post-intervention	7.52 (5.25–9.78)	6.26 (4.11–8.40)	9.10 (6.35–11.84)
	follow-up	7.42 (5.15–9.68)	5.60 (3.46–7.75)	6.82 (4.08–9.57)

Dose Response

To determine the dose response of yoga activity on PTSD compared to STC, hierarchical regression was conducted in which predictors were added based on theoretical importance while checking for multicollinearity. A series of models were built, first controlling for depression, BMI, age, household income, education level, race, and peer support. Marital status was not included due to lack of variability in STC. Due to multicollinearity determined from significant correlations ($p < 0.05$) with BMI, household income ($r = -.41$, $p = 0.009$), education level ($r = -.41$, $p = 0.009$), and race ($r = .36$, $p = 0.021$) were not included in the final model, which controlled for BMI, age, and peer support. Participating in yoga was a non-significant predictor of reduced IES-R scores (t (32) = −0.50, β=−0.031, 95% CI [−0.196, 0.095], $p = 0.621$). contradictory to participating in the STC group t (12) = 0.121, β=0.046, 95% CI [−0.924, 1.016], $p = 0.908$). For participants in the LD or MD groups, 10-min increases in yoga activity weekly for 12 weeks (i.e. 120 min increase averaged over 12 weeks) resulted in a 3.72 unit decrease in PTSD whereas the STC resulted in a 5.52 unit increase in PTSD.

DISCUSSION

The purpose of this study was to determine the feasibility and acceptability of a 12-week, home-based, online-streamed yoga intervention with varying doses (low= 60 min/week; moderate= 150 min/week) among bereaved mothers who had experienced the death of a baby to stillbirth. This was the first randomized controlled study to determine the feasibility of online yoga in grieving mothers. Although just under half the participants did not finish the study, many of those who dropped out mentioned that they enjoyed the online videos (almost 64%) and thought social support online would have helped them complete the study (55%). Overall participants were satisfied with the intervention and thought it affected them in a positive way. Only a small number of participants met the benchmarks for demand, yet there were significant reductions in depression and perinatal grief in the MD group as compared to the STC. Additionally, there were notable non-significant reductions in PTSD symptoms and improvements in sleep quality. With both intervention groups showing similar trends for improvement, these preliminary efficacy data also suggest that lower doses of yoga may be enough to elicit reductions in post traumatic stress symptoms.

Acceptability

Just under 46% of participants did not finish the study (i.e., dropped out or did not complete post), mainly because of time, mood (e.g., lack of energy and motivation), and stress. It was not surprising that time was a major reason for not completing the study as other physical activity interventions have also revealed time as a barrier, especially in pregnant and post-partum populations and women in the inter-conception period [45, 46]. These studies report that childcare and family obligations consume the time that they might have for physical activity/exercise [46]. Women also reported mood as a reason for not completing the study. Participants in our study were within 2 years of experiencing a baby's death. Some literature suggests symptoms related to post traumatic stress are most prevalent and severe in the first year following the loss, and these symptoms interfere with normal life activities [19]. Mood-related symptoms associated with PTSD may have inhibited motivation to complete the prescribed videos each week. Although the prevalence and patterns of mood-related symptoms after stillbirth are not known, one study found that women diagnosed with PTSD after live-birth had higher levels of anxiety, depression, and parenting stress through 12 months postpartum [47]. The time since the baby's death could play a key role in the ability of a woman to participate in inter-conception care (e.g., delivery of intervention based on time since the stillbirth) [48], and this information could inform the design of future interventions. Additionally, individuals experiencing symptoms of post-traumatic stress may perceive everyday stressors as more severe than prior to the traumatic event and may not have proper coping mechanisms to handle minor stressors [49]. If the intervention was perceived as an added stressor (e.g., another item on a to-do list), participants may have not been motivated to participate. That is, they may be more inclined to avoid these stressors until they have developed coping mechanisms to manage stressors.

Interestingly, participants who dropped out of the study reported they enjoyed participating in online yoga or exercise and thought if the study included an online social support group they would have been more likely to continue. Social support is a well-known behavior strategy to increasing compliance and attendance to behavioral interventions [50] Social support is critical for women after stillbirth as many report feeling isolated and their grief is often disenfranchised [51]. It is a key component in coping with loss and trauma, and those who report lower levels of social support have reported more adverse health outcomes [52] Social support may also play a role in the effects of yoga on mental health outcomes as it has been reported that social support provided during classes may be a coping mechanism for stress [53] Because the kind of support that would naturally occur in an in-person class was not present in this online intervention, there is a need to determine feasible, alternative support strategies.

Satisfaction

Overall, participants were satisfied with the intervention, reporting they enjoyed the intervention and instructors. Similar to those who dropped out of the intervention, most participants reported they would have participated more if there had been an online social support group. Social support has been used in various physical activity interventions in order to increase adherence and enjoyment, and in some populations has been thought to mediate the effects of the intervention [54].

In the participants who became pregnant during the study and completed the intervention, many reported discomfort and being afraid of hurting the baby as barriers to participation. These women also did not have the opportunity to receive advice for modifications as they would have in an in-person prenatal yoga class. In order to progress the feasibility of online yoga interventions, future studies should consider including an online mode of delivery for a yoga instructor to provide feedback and assistance in order to prevent injury and offer modifications. For example, participants could use a real-time video software to engage with an instructor while actually performing the poses. Yoga instructors may offer support and a feeling of safety, ensure proper techniques are employed, and increase motivation.

Another obstacle with online yoga is the potential for technical issues. Most participants who reported experiencing technical issues said it was due to slow internet connections. One way to address this in future studies might be to provide instruction and recommendations for the participants to download the videos before watching them to complete their sessions.

When asked what barriers kept them from participating in yoga (on a weekly basis), women reported similar barriers to the women that dropped out of the study (i.e., work, caring for others, mood, time). In our study, we provided a barriers workbook (i.e., tips for overcoming barriers to participation) and weekly reminders to participate. The literature is not conclusive regarding overcoming barriers to exercise, but reminders seem to be effective for some populations [55]. Because of the high dropout rate and number of women who reported barriers, there is a need to develop strategies for fidelity to an online yoga intervention. We utilized reminders to participate in yoga or STC but did not include any behavioral strategies or language to overcome barriers within these reminders. Future studies could include reminders with tips that are specific to women who have had a stillbirth (i.e., sensitivity to grief, awareness of other responsibilities such as children). Additionally, because social support is necessary for improving health outcomes after a stillbirth, future research may utilize social support as a means to overcome barriers. For instance, women in the study could connect with other women who are participating and discuss ideas for overcoming their barriers such as scheduling time and childcare, and provide emotional support for each other.

Demand

Most women in the MD group reported 150 min per week to be too much and, not surprisingly, did not meet the benchmarks. Interestingly, although most participants in the low dose group reported a dose of 60 min per week of yoga to be appropriate, less than half of participants met the benchmarks for participation. Low fidelity is not surprising considering the high distress within this population [56] Other studies in populations with PTSD symptoms have reported similar compliance rates [57] Schottenbauer and colleagues found that it was common for dropout rates to be around 50% in high posttraumatic stress populations [58]. Reinhardt and colleagues also reported low retention rates in a yoga intervention for veterans with PTSD [57]. In a yoga intervention for survivors of domestic violence, Clark and colleagues reported a 30% dropout rate [59]. More research is needed to determine intervention strategies to increase fidelity specific to this population.

Preliminary Efficacy

Because most participants did not meet the benchmarks for minutes of yoga participation per week, preliminary efficacy results are interpreted with average minutes of actual participation per week (LD=44, MD=77) rather than prescribed minutes per week (LD=60, MD=150). There was a significant reduction in posttraumatic stress and depressive symptoms at post-intervention but not at follow-up. Other studies have shown short-term effects of yoga on depressive and PTSD symptoms (e.g., 20) but more research is needed to determine the long-term effects of yoga on these outcomes. The physical domain of self-rated health decreased over the course of the intervention. Our findings are inconsistent to self-rated health outcomes in yoga interventions overall [60]. However, women who have experienced a stillbirth report physical health declines and may be more likely to develop chronic illness [61]. Although women in the STC and LD groups experienced decreases in their physical self-rated health, women in the MD group did not have any changes. This suggests yoga may have mitigated the typical decline in self-rated health seen in mothers who experience stillbirth. More research is needed in this population, specifically, to determine implications of self-rated health as it relates to post traumatic stress and other negative psychological and physical health outcomes.

The significant decreases in depression and perinatal grief in the MD group compared to the STC group suggests that 77 min of yoga per week may be enough to improve these outcomes. Other studies have shown similar improvements in depressive symptoms with 60–75 min of yoga per week compared to control groups (e.g., [62, 63]) It should be noted that most yoga interventions have been conducted using in-person formats as opposed to the online, home-based format in the current study. Therefore, discrepancies between number and length of sessions are inherent despite the potentially identical number of minutes per week. More research is needed to determine the differences and efficacy between the two methods. Additionally, although yoga may decrease depressive symptoms and grief, both states may also be barriers to participation and less likely to participate [64].

Although not significant, trends in improvements of post traumatic stress symptoms, perinatal grief, and sleep quality in both LD and MD groups, compared to the STC group, also suggest the intervention may contribute to these changes regardless of dose, a concept further supported when additionally considering regression results. Decreases on the IES-R in both intervention groups was the largest relative change among all outcome measures, suggesting a potential for yoga to decrease symptoms at doses of both 44 and 77 min.

Our dose-response findings, while also non-significant, suggested potential decreases in post traumatic stress and depression scores for every 10-min increase of yoga per week. Taken together, these trends suggest low doses of yoga (approximately 20 min on average) may be enough to elicit positive changes but more minutes per week may yield greater magnitudes of improvement. That said, these findings should also be considered in light of the feasibility findings previously discussed. That is, even though there may be an increase in benefits with additional participation, participants reported

60 min to be feasible whereas 150 min was too much. More research is needed to determine the point at which benefits are balanced with feasibility.

Limitations

This was the first study to determine the feasibility of online yoga in women who have experienced stillbirth and therefore had limitations, many related to the online nature of the intervention. First, there were inconsistencies with the software used to track participation (i.e., viewing the videos) and data for some participants was not accurately recorded. Therefore, the research team utilized self-report data and the results were subject to biases that are inevitable with this type of data (e.g., social desirability). In the satisfaction surveys, participants reported technical issues that limited viewing the videos (e.g., slow internet) and may have inhibited their participation and the ability to track the minutes appropriately. Future studies should consider ways to overcome such issues that are likely inherent with online interventions. Second, the sample was not necessarily representative of the population of the women who have had stillbirths in the US, as most were white and higher income. Although in a previous study we asked minority women who had experienced stillbirth about cultural barriers to recruitment and about their perceptions and opinions on yoga [18], additional recruiting methods are needed for representative samples to ensure generalizability of the results. Third, this study had a high number of dropouts. Participants had within the last 2 years, experienced a traumatic event, and may be coping with a number of stressors. Future studies should consider strategies recommended above (i.e., social support) to increase retention and adherence to an online yoga intervention. The nature of social media recruitment may have led to a self-selection bias in this study. Finally, this was a feasibility study and thus our sample size was not powered to determine efficacy. That is, conclusions cannot be made about the effect of yoga on any of the preliminary outcome measures (e.g., PTSD, anxiety, depression). Though findings showed trends in the hypothesized directions, future studies warrant sample sizes powered to determine efficacy.

CONCLUSION

Online yoga may be an effective and feasible strategy to reduce depressive symptoms, perinatal grief, and symptoms of PTSD in mothers who have experienced stillbirth. More research is needed to determine strategies to increase compliance, the most appropriate number of minutes per week in order to experience changes in outcomes, and the effects of the intervention compared to a control group.

AVAILABILITY OF DATA AND MATERIALS

Available from the corresponding author upon reasonable request.

ABBREVIATIONS

PTSD:	Post-traumatic stress disorder
LD:	Low dose intervention group
MD:	Moderate dose intervention group
STC:	Stretch and tone control group
IES-R:	Impact of Event Scale-Revised
STAI:	State-Trait Anxiety Inventory
PHQ-9:	Patient Health Questionnaire-9
PGS:	Perinatal Grief Scale
SCS:	Self-Compassion Scale
ERQ:	Emotion Regulation Questionnaire
MAAS:	Mindful Attention Awareness Scale
PSQI:	Pittsburgh Sleep Quality Index
SF-12:	Self-Rated Health Survey-Short Form

REFERENCES

1. Pollock D, Ziaian T, Pearson E, Cooper M, Warland J. Understanding stillbirth stigma: a scoping literature review. Women Birth. 2019;3(33):207-18.
2. Hoyert DL, Gregory EC. Cause of fetal death: data from the fetal death report, 2014. Natl Vital Stat Rep. 2016;65(7):1–25.
3. Xu J, Koachaenek K, Murphy S, Arias E. Mortality in the United States, 2012. 2014. Contract No: Report.
4. Gold KJ, Leon I, Boggs ME, Sen A. Depression and posttraumatic stress symptoms after perinatal loss in a population-based sample. J Womens Health. 2016;25(3):263–9.
5. Hutti MH, Armstrong DS, Myers JA, Hall LA. Grief intensity, psychological well-being, and the intimate partner relationship in the subsequent pregnancy after a perinatal loss. J Obstet Gynecol Neonatal Nurs. 2015;44(1):42–50.
6. Côté A, Daneault S. Effect of yoga on patients with cancer our current understanding. Can Fam Physician. 2012;58(9):e475–E9.
7. Gaudet CC. Pregnancy after perinatal loss: association of grief, anxiety and attachment. J Reproductive Infant Psychol. 2010;28(3):240 –51; 51.
8. Turton P, Evans C, Hughes P. Long-term psychosocial sequelae of stillbirth: phase II of a nested case-control cohort study. Archives Women's Mental Health. 2009;12(1):35–41.
9. Gravensteen IK, Jacobsen EM, Sandset PM, Helgadottir LB, Rådestad I, Sandvik L, et al. Anxiety, depression and relationship satisfaction in the pregnancy following stillbirth and after the birth of a live-born baby: a prospective study. BMC Pregnancy Childbirth. 2018;18(1):41.
10. Huberty JL, Matthews J, Leiferman J, Hermer J, Cacciatore J. When a baby dies: a systematic review of experimental interventions for women after stillbirth. Reprod Sci. 2017;24(7):967–75.
11. Lake MF, Johnson TM, Murphy J, Knuppel RA. Evaluation of a perinatal grief support team. Am J Obstet Gynecol. 1987;157(5):1203–6.
12. Roberts LR, Montgomery SB. Mindfulness-based intervention for perinatal grief after stillbirth in rural India. Issues Mental Health Nursing. 2015;36(3):222–30.

13. Ellis A, Chebsey C, Storey C, Bradley S, Jackson S, Flenady V, et al. Systematic review to understand and improve care after stillbirth: a review of parents' and healthcare professionals' experiences. BMC Pregnancy Childbirth. 2016;16(17):16.
14. Iyengar BKS. BKS Iyengar yoga: the path to holistic health: penguin; 2007.
15. Nolan CR. Bending without breaking: a narrative review of trauma-sensitive yoga for women with PTSD. Complement Ther Clin Pract. 2016;24:32–40.
16. Macy RJ, Jones E, Graham LM, Roach L. Yoga for trauma and related mental health problems: a meta-review with clinical and service recommendations. Trauma Violence Abuse. 2018;19(1):35–57.
17. Buttner MM, Brock RL, O'Hara MW, Stuart S. Efficacy of yoga for depressed postpartum women: a randomized controlled trial. Complement Ther Clin Pract. 2015;21(2):94–100.
18. Huberty J, Green J, Gold KJ, Leiferman J, Cacciatore J. An iterative design process to develop a randomized feasibility study and inform recruitment of minority women after stillbirth. Pilot Feasibility Studies. 2019;5(1):140.
19. Hollins Martin CJ, Martin CR. Post-traumatic stress disorder (PTSD) and its relationship with perinatal bereavement: definitions, reactions, adjustments, and grief. Comprehensive Guide to Post-Traumatic Stress Disorders. 2016:599–626. https://doi.org/10.1007/978-3-319-08359-9_44.
20. Mitchell KS, Dick AM, DiMartino DM, Smith BN, Niles B, Koenen KC, et al. A pilot study of a randomized controlled trial of yoga as an intervention for PTSD symptoms in women. J Trauma Stress. 2014;27(2):121–8.
21. Uebelacker LA, Tremont G, Epstein-Lubow G, Gaudiano BA, Gillette T, Kalibatseva Z, et al. Open trial of Vinyasa yoga for persistently depressed individuals: evidence of feasibility and acceptability. Behav Modif. 2010;34(3):247–64.
22. Khalsa SB. Yoga as a therapeutic intervention: a bibliometric analysis of published research studies. Indian J Physiol Pharmacol. 2004;48(3):269–85.
23. Jain S, Shapiro SL, Swanick S, Roesch SC, Mills PJ, Bell I, et al. A randomized controlled trial of mindfulness meditation versus relaxation training: effects on distress, positive states of mind, rumination, and distraction. Ann Behav Med. 2007;33(1):11–21.

24. Hölzel BK, Lazar SW, Gard T, Schuman-Olivier Z, Vago DR, Ott U. How does mindfulness meditation work? Proposing mechanisms of action from a conceptual and neural perspective. Perspect Psychol Sci. 2011;6(6):537–59.

25. Halpern J, Cohen M, Kennedy G, Reece J, Cahan C, Baharav A. Yoga for improving sleep quality and quality of life for older adults. Altern Ther Health Med. 2014;20(3):37–46.

26. Beddoe AE, Lee KA, Weiss SJ, Kennedy HP, Yang CP. Effects of mindful yoga on sleep in pregnant women: a pilot study. Biol Res Nurs. 2010;11(4):363–70.

27. Staples JK, Hamilton MF, Uddo M. A yoga program for the symptoms of post-traumatic stress disorder in veterans. Mil Med. 2013;178(8):854–60.

28. Cacciatore J, editor Psychological effects of stillbirth. Seminars in Fetal and Neonatal Medicine; 2013: Elsevier. https://doi.org/10.1016/j.siny.2012.09.001.

29. Wistia Inc. Cambridge, MA. 2008.

30. McAuley E, Wójcicki TR, Gothe NP, Mailey EL, Szabo AN, Fanning J, et al. Effects of a DVD-delivered exercise intervention on physical function in older adults. J Gerontol A Biol Sci Med Sci. 2013;68(9):1076–82.

31. Bowen DJ, Kreuter M, Spring B, Cofta-Woerpel L, Linnan L, Weiner D, et al. How we design feasibility studies. Am J Prev Med. 2009;36(5):452–7.

32. Creamer M, Bell R, Failla S. Psychometric properties of the impact of event scale—revised. Behav Res Ther. 2003;41(12):1489–96.

33. Newham JJ, Westwood M, Aplin JD, Wittkowski A. State-trait anxiety inventory (STAI) scores during pregnancy following intervention with complementary therapies. J Affect Disord. 2012;142(1–3):22–30.

34. Yawn BP, Pace W, Wollan PC, Bertram S, Kurland M, Graham D, et al. Concordance of Edinburgh postnatal depression scale (EPDS) and patient health questionnaire (PHQ-9) to assess increased risk of depression among postpartum women. J Am Board Fam Med. 2009;22(5):483–91.

35. Sidebottom AC, Harrison PA, Godecker A, Kim H. Validation of the patient health questionnaire (PHQ)-9 for prenatal depression screening. Arch Womens Ment Health. 2012;15(5):367–74.

36. Toedter LJ, Lasker JN, Alhadeff JM. The perinatal grief scale: development and initial validation. Am J Orthopsychiatry. 1988;58(3):435–49.

37. **CAS** Neff KD. Development and validation of a scale to measure self-compassion. Self Iden. 2003;2:223.

38. Gross JJ, John OP. Individual differences in two emotion regulation processes: implications for affect, relationships, and well-being. J Pers Soc Psychol. 2003;85(2):348–62.

39. Gratz KL, Roemer L. Multidimensional assessment of emotion regulation and dysregulation: development, factor structure, and initial validation of the difficulties in emotion regulation scale. J Psychopathol Behav Assess. 2004;26(1):41–54.

40. MacKillop J, Anderson EJ. Further psychometric validation of the mindful attention awareness scale (MAAS). J Psychopathol Behav Assess. 2007;29(4):289–93.

41. Mindell JA, Cook RA, Nikolovski J. Sleep patterns and sleep disturbances across pregnancy. Sleep Med. 2015;16(4):483–8.

42. Turner-Bowker D, Hogue SJ. Short form 12 health survey (SF-12). Encyclopedia Quality Life Well-Being Research. 2014:5954–7.

43. Field A. Discovering statistics using IBM SPSS statistics: north American edition: sage; 2017.

44. Spss I. Statistics for windows, version 24. 0 [computer software]. Armonk: IBM Corp.; 2016.

45. Saligheh M, McNamara B, Rooney R. Perceived barriers and enablers of physical activity in postpartum women: a qualitative approach. BMC Pregnancy Childbirth. 2016;16(1):131.

46. Mailey EL, Huberty J, Dinkel D, McAuley E. Physical activity barriers and facilitators among working mothers and fathers. BMC Public Health. 2014;14:657–2458–14-657.

47. Goulding A, Grewen K, Meltzer-Brody S, Pearson B, Stuebe A. Maternal perception of birth trauma and its association with postpartum mood and parenting stress. American J Obstetrics Gynecol. 2019;220(1):S173.

48. Badura M, Johnson K, Hench K, Reyes M. Healthy start lessons learned on interconception care. Womens Health Issues. 2008;18(6 Suppl):S61–6.

49. Laganà L, Reger SL. A pilot study on perceived stress and PTSD symptomatology in relation to four dimensions of older women's physical health. Aging Ment Health. 2009;13(6):885–93.

50. DiMatteo MR. Social support and patient adherence to medical treatment: a meta-analysis. Health Psychol. 2004;23(2):207–18.

51. Cacciatore J, Schnebly S, Froen JF. The effects of social support on maternal anxiety and depression after stillbirth. Health Soc Care Community. 2009;17(2):167–76.

52. Mekosh-Rosenbaum V, Lasker JN. Effects of pregnancy outcomes on marital satisfaction: a longitudinal study of birth and loss. Infant Ment Health J. 1995;16(2):127–43.

53. Evans S, Tsao JC, Sternlieb B, Zeltzer LK. Using the biopsychosocial model to understand the health benefits of yoga. Journal of complementary and integrative medicine. 2009;6(1):1-22.

54. Klaben ME, McDonough MH, Daun JT, Beselt LJ, Culos-Reed SN, Kronlund LJ, et al. The effect of social support in physical activity interventions on physical activity and quality of life among cancer survivors: A systematic review of randomized control trials and quasi-experimental studies. J Exercise, Movement Sport (SCAPPS refereed abstracts repository). 2017;49(1):176.

55. Wanner M, Martin-Diener E, Bauer G, Braun-Fahrländer C, Martin BW. Comparison of trial participants and open access users of a web-based physical activity intervention regarding adherence, attrition, and repeated participation. J Med Internet Res. 2010;12(1):e3.

56. 5Turton P, Hughes P, Evans CD, Fainman D. Incidence, correlates and predictors of post-traumatic stress disorder in the pregnancy after stillbirth. Brit J Psychiatry. 2001;178:556–60.

57. Reinhardt KM, Noggle Taylor JJ, Johnston J, Zameer A, Cheema S, Khalsa SBS. Kripalu yoga for military veterans with PTSD: a randomized trial. J Clin Psychol. 2018;74(1):93–108.

58. Schottenbauer MA, Glass CR, Arnkoff DB, Tendick V, Gray SH. Nonresponse and dropout rates in outcome studies on PTSD: review and methodological considerations. Psychiatry. 2008;71(2):134–68.

59. Clark CJ, Lewis-Dmello A, Anders D, Parsons A, Nguyen-Feng V, Henn L, et al. Trauma-sensitive yoga as an adjunct mental health treatment in group therapy for survivors of domestic violence: a feasibility study. Complement Ther Clin Pract. 2014;20(3):152–8.

60. Benavidez G, Hart PD. Effects of yoga on measures of health-related quality of life from sf-36 and sf-12 assessments: a systematic review and meta-analysis. Exercise Medicine. 2017;1:1-11.
61. Burden C, Bradley S, Storey C, Ellis A, Heazell AEP, Downe S, et al. From grief, guilt pain and stigma to hope and pride - a systematic review and meta-analysis of mixed-method research of the psychosocial impact of stillbirth. BMC Pregnancy Childbirth. 2016;16(1):9.
62. Dick AM, Niles BL, Street AE, DiMartino DM, Mitchell KS. Examining mechanisms of change in a yoga intervention for women: the influence of mindfulness, psychological flexibility, and emotion regulation on PTSD symptoms. J Clin Psychol. 2014;70(12):1170–82.
63. van der Kolk BA, Stone L, West J, Rhodes A, Emerson D, Suvak M, et al. Original research yoga as an adjunctive treatment for posttraumatic stress disorder: a randomized controlled trial. J Clin Psychiatry. 2014;75(6):e559–e65.
64. Moadel AB, Shah C, Wylie-Rosett J, Harris MS, Patel SR, Hall CB, et al. Randomized controlled trial of yoga among a multiethnic sample of breast cancer patients: effects on quality of life. J Clin Oncol. 2007;25(28):4387–95.

Chapter 7

The Efficacy of a Comprehensive Lifestyle Modification Programme Based on Yoga in the Management of Bronchial Asthma: A Randomized Controlled Trial

Ramaprabhu Vempati[1], Ramesh Lal Bijlani[1,2] and Kishore Kumar Deepak[1]

[1]Department of Physiology, All India Institute of Medical Sciences, New Delhi, India

[2]Sri Aurobindo Ashram, New Delhi, India

ABSTRACT

Background

There is a substantial body of evidence on the efficacy of yoga in the management of bronchial asthma. Many studies have reported, as the effects of yoga on bronchial asthma, significant improvements in pulmonary functions, quality of life and reduction in airway hyper-reactivity, frequency of attacks and medication use. In addition, a few studies have attempted

Citation: The efficacy of a comprehensive lifestyle modification programme based on yoga in the management of bronchial asthma: a randomized controlled trial Ramaprabhu Vempati, Ramesh Lal Bijlani and Kishore Kumar Deepak. BMC Pulmonary Medicine 2009, 9:37 doi:10.1186/1471-2466-9-37.

to understand the effects of yoga on exercise-induced bronchoconstriction (EIB) or exercise tolerance capacity. However, none of these studies has investigated any immunological mechanisms by which yoga improves these variables in bronchial asthma.

Methods

The present randomized controlled trial (RCT) was conducted on 57 adult subjects with mild or moderate bronchial asthma who were allocated randomly to either the yoga (intervention) group (n = 29) or the wait-listed control group (n = 28). The control group received only conventional care and the yoga group received an intervention based on yoga, in addition to the conventional care. The intervention consisted of 2-wk supervised training in lifestyle modification and stress management based on yoga followed by closely monitored continuation of the practices at home for 6-wk. The outcome measures were assessed in both the groups at 0 wk (baseline), 2, 4 and 8 wk by using Generalized Linear Model (GLM) repeated measures followed by post-hoc analysis.

Results

In the yoga group, there was a steady and progressive improvement in pulmonary function, the change being statistically significant in case of the first second of forced expiratory volume (FEV_1) at 8 wk, and peak expiratory flow rate (PEFR) at 2, 4 and 8 wk as compared to the corresponding baseline values. There was a significant reduction in EIB in the yoga group. However, there was no corresponding reduction in the urinary prostaglandin D_2 metabolite (11β prostaglandin F2α) levels in response to the exercise challenge. There was also no significant change in serum eosinophilic cationic protein levels during the 8-wk study period in either group. There was a significant improvement in Asthma Quality of Life (AQOL) scores in both groups over the 8-wk study period. But the improvement was achieved earlier and was more complete in the yoga group. The number-needed-to-treat worked out to be 1.82 for the total AQOL score. An improvement in total AQOL score was greater than the minimal important difference and the same outcome was achieved for the sub-domains of the AQOL. The frequency of rescue medication use showed a significant decrease over the study period in both the groups. However, the decrease was achieved relatively earlier and was more marked in the yoga group than in the control group.

Conclusion

The present RCT has demonstrated that adding the mind-body approach of yoga to the predominantly physical approach of conventional care results in measurable improvement in subjective as well as objective outcomes in bronchial asthma. The trial supports the efficacy of yoga in the management of bronchial asthma. However, the preliminary efforts made towards working out the mechanism of action of the intervention have not thrown much light on how yoga works in bronchial asthma.

Trial registration

Current Controlled Trials ISRCTN00815962

BACKGROUND

Although the efficacy of yoga in treating bronchial asthma[1] has been investigated since at least the 1960s [2–9], most of the previous studies have been uncontrolled, and have evaluated only a few selected yogic postures or breathing exercises. To the best of our knowledge, there is only one randomized controlled trial which has evaluated the efficacy of an integrated package consisting of yogic postures, breathing exercises, cleansing techniques, meditation, devotional sessions and lectures[10]. However, even this study did not investigate the mechanisms by which yoga improves the symptoms of bronchial asthma. The present randomized controlled trial was undertaken to study the efficacy of a comprehensive lifestyle modification and stress management program based on yoga in subjects having mild or moderate bronchial asthma. An attempt has also been made to monitor some immunological indicators of severity of disease and mast cell activation.

METHODS

Subjects

The subjects were adult patients having mild or moderate bronchial asthma who were either referred to the Integral Health Clinic (IHC) of the All India Institute of Medical Sciences (AIIMS) by AIIMS physicians or came to IHC in response to our advertisements in local dailies.

The potential subjects went through a step-wise screening procedure. The inclusion criteria consisted of (1) age 18 years or older; (2) an established diagnosis of mild-to-moderate asthma for at least 6 months (meeting the American Thoracic Society[11] spirometry criteria for mild-to-moderate asthma, which requires either a forced expiratory volume in 1 second [FEV_1]/forced vital capacity [FVC] below the lower limit of normal with a significant response to a bronchodilator [a ≥12% increase and a ≥200 mL absolute increase in FEV_1 15 minutes after the administration of 2 puffs of a short acting β-agonist] or peak expiratory flow rate [PEFR] variability >20%); (3) taking at least one of the following: inhaled β-agonists, methylxanthines, anticholinergics, inhaled corticosteroids; and (4) stable medication dosing for the past month. Subjects were excluded if they (1) smoked currently (or in the past year) or had a smoking history of greater than 5 pack-years; (2) had a concomitant lung disease; (3) were taking leukotriene inhibitors or receptor antagonists, or mast cell-stabilizing agents for at least 6 months; (4) practiced yoga or any other similar discipline during 6 months preceding the study; (5) were pregnant; (6) had a chronic medical condition that required treatment with oral or systemic corticosteroids in the past month; (7) had a medical condition that contraindicated exercise; or (8) had an unstable medical condition.

Of the 138 potential subjects screened, 66 did not meet the above-described eligibility criteria. Of the 72 eligible subjects, only 60 were included in the study at the end of the 1-week run-in period and the other 12 subjects were self-withdrawn before randomization. The subjects completed the baseline measurements of all the outcome parameters after randomization. These 60 subjects were randomly allocated to either the yoga (intervention) group (n = 30) or the wait-listed control group (n = 30). However, one subject in the yoga group, and two subjects in the control group discontinued midway in the study. The results presented in this report are based on the data collected from only the 57 subjects who completed the study (yoga group, n = 29; control group, n = 28).

Experimental Design

The present study is an open-label randomized controlled trial (RCT) with two parallel groups. The control group received conventional care whereas the experimental group received, in addition to conventional care, an intervention based on yoga. The subjects from both groups were on bronchodilators in the form of inhalers or inhaled corticosteroids depending on the judgment of the treating physician. None of the subjects was on systemic or oral corticosteroid therapy. However, subjects were not allowed to change their inhaled corticosteroid dose during the study period. During the study, the control group was offered a session on health education relevant to their illness, and at the end of the 8-wk study period, the intervention based on yoga. The outcome measures were assessed in both groups at 0 wk (baseline), 2, 4 and 8 wk. The subjects in neither group were encouraged to either delay or reduce their rescue medication use, but were instead advised to consult their treating physician for any dose changes, if required.

Treatment Intervention

Participants assigned to the yoga group underwent a comprehensive yoga-based lifestyle modification and stress management program for 4 hours a day for 2 weeks. The sessions were conducted from 8 am to 12 noon. The subjects were neither unemployed nor disabled. Their ability to spend that much time in the clinic might be due to 'cultural differences'. In India, the patient respects his therapist and is willing to do whatever it takes to comply with the prescribed treatment. Corresponding to this attitude, the employer is also willing to bend the requirements of the job to make it possible for the employee to get the treatment. If the patient is a homemaker, other members of the extended family are willing to pitch in with childcare and other essential tasks to enable the patient to get the treatment. The program consisted of lectures and practical sessions on asanas (postures), pranayamas (breathing techniques), kriyas (cleansing techniques), meditation and shavasana (a relaxation technique). The 4-hour sessions included asanas and pranayamas for 1 hour; breakfast and building up of group support for 30 min; lecture and discussion for 2 hours and meditation for 30 min. The yoga practice sessions were conducted by a qualified yoga instructor. The lectures were on yoga, its place in daily life, its application to stress management, fundamentals of nutrition and health education relevant to their illness. Each subject received at least one session of about one hour of individualized counseling. During these individualized counselling sessions, the subject's questions regarding yogic practices, diet, prognosis, test results etc. were answered. In addition, family and social histories of the subjects were taken to establish a good rapport, and to discover any major mental stress, real or perceived, which might be aggravating the disease. In case of mental stress, some counselling, primarily in the form of cognitive restructuring based on the spiritual philosophy underlying yoga, was attempted. The counselling was done by physicians with special interest in yoga and mind-body medicine.

The physical practices consisted of elementary breathing exercises and loosening exercises followed by asanas under four categories (standing, sitting, prone and supine), pranayama, kriyas, and meditation based on raja yoga. Participants were instructed to follow a specific breathing pattern during each asana and asked to hold each pose in final position for up to 30 sec without holding their breath. Soothing instrumental music was played while the participants were practicing the yoga techniques to induce relaxation. Participants practiced yogic postures with awareness focused on their physical movements and breath. Each posture was followed by an appropriate relaxation technique for a short period. The classes concluded with deep relaxation in shavasana for about 15 min followed by guided imagery for about 5 min. The yoga practice sessions were about one and a half hour during the 2-wk training period and additional 6-wk home practice. The level of compliance for yoga practice was assured at least 5 days a week. The content of the daily home practice sessions was the same as during the 2-wk training, i.e. one hour of asanas and pranayama, 10 min of relaxation and 20 min of meditation. Subjects were provided printed material and audiocassettes to supplement live instruction. The yoga group subjects' compliance during the follow-up period was monitored by a diary on their adherence to yoga practice, dietary advice and rescue medication use, which they filled in everyday and brought during each visit. An individual yoga practice session was offered to the subjects during study evaluation visits. The control group was also asked to keep a diary of medication use and was attended by their treating physicians during study evaluation visits. During the follow-up period, telephonic support was also provided for motivating participants to improve their compliance.

The diet recommended for the yoga group was predominantly vegetarian; it mainly consisted of a combination of cereals and pulses, preferably unrefined; moderate amounts of judiciously chosen fat; moderate amounts of milk and milk products; moderate amounts of spices; vegetables and fruits about 500 g, vegetables being predominantly of the leafy green variety, and at least some eaten raw (uncooked).

Outcome Measures and Techniques of Measurement

The outcome measures were spirometric indices of pulmonary function, eosinophilic cationic protein as a marker of the course of the disease[12]; the degree of exercise-induced bronchoconstriction, urinary concentration of the prostaglandin D_2 metabolite, 11β prostaglandin F2α before and after the exercise challenge as a marker of mast cell activation, quality of life and frequency of rescue medication.

Spirometric Indices of Pulmonary Function

The indices of interest, viz. forced vital capacity (FVC), forced expiratory volume in the first second of expiration (FEV_1), the ratio FEV_1/FVC, peak expiratory flow rate (PEFR) and average forced expiratory flow rate during the expulsion of 25–75% of FVC ($FEF_{25-75\%}$) were measured by computerized spirometric equipment (K4b2, COSMED srl, Italy). The FVC manoeuvre was repeated at least thrice during each measurement. The highest of three acceptable readings of FVC and FEV_1 were taken as the values for that sitting.

Eosinophilic Cationic Protein (ECP)

The serum concentration of ECP was assessed by sandwich ELISA kit (MBL ECP ELISA kit, Japan, code No. 7618 E). The coefficients of variation for inter-assay and intra-assay measurements were 10.7% and 11.1% respectively.

Exercise-Induced Bronchoconstriction (EIB)

The exercise challenge was given using a stationary bicycle ergometer (Bodyguard® 990, Sandnes, Norway). The workload was fixed at 80–85% of the maximum predicted heart rate while pedalling at 60 revolutions per min. The subjects were told to go on till they reach the limit of their dyspnoea or inability to continue the exercise for any reason. The duration of the exercise in different subjects ranged from 3–7 min. The heart rate was monitored continuously through a remote sensor attached to the subject's chest (Polar Electro Inc., Finland) and displayed on a computer screen. The exercise was performed in an air-conditioned room with ambient temperature between 24 and 27°C, and relative humidity between 40 and 50%. Thus the environmental conditions under which exercise was performed at different sittings were comparable. The degree of EIB was measured as highest degree of exercise-induced change in FEV_1 from 0 min (baseline) to 3, 8, 15 and 30 min post exercise.

Urinary 11β prostaglandin F2α (11β-PGF2α)

Urine was collected 5 min before exercise and again 30 min after the end of exercise. Urine samples were stored, without the addition of any preservatives, at -20°C until analysis. The concentration of 11β-PGF2α excretion was estimated by using Enzyme Immuno Assay (EIA, Cayman

Chemical, Ann Arbor, MI, USA; Catalog No. 516521). The inter-assay and intra-assay variations for 11β-PGF2α assay were 9.7% and 5.5% respectively. To correct for the effect of diuresis, creatinine concentrations were measured for all urine samples by a commercially available colorimetric assay (Wipro Biomed, Bangalore, India), using Jaffe' Rate method[13]. Therefore, the urinary 11β prostaglandin F2α concentrations were expressed as picogram per milligram of creatinine[14].

Quality of Life

Quality of life was measured by using a self-administered Asthma Quality of Life Questionnaire (AQOL) which is available in bilingual form, i.e. English and Hindi[15]. The AQOL is a 32-item disease specific questionnaire that has been validated to measure the problems that adult patients with asthma experience in their daily lives[16] and has also been validated on Indian population[17].

Subjects responded to each question on a 7-point scale (1 being maximum impairment; 7 being no impairment). The overall quality of life score is the mean score of all the 32 items. Thus the score may vary from 1–7. The 32 items are further grouped into four sub-domains viz., symptoms, activity limitation, emotional function and reactivity to environmental stimuli. The score for each sub-domain was also calculated as the mean score for items pertaining to that sub-domain. Thus, the score for each sub-domain may also vary from 1–7.

Rescue Medication

The subjects were asked at the beginning of the study to add up how many times they had used a bronchodilator inhaler during the preceding 2-wk as prescribed by the physicians. For quantifying subsequent use, the subjects were told to make a note in their diaries every time they had to use rescue medication for getting quick relief from symptoms such as wheezing, cough or shortness of breath. At any point in time, the frequency of rescue medication was calculated as the average number of times that the medication had to be used in a day during the preceding 2-wk.

Analysis of Data

The baseline (0 wk) values of each outcome measure were compared with the values obtained at 2, 4 and 8 wk by Generalized Linear Model (GLM) repeated measures followed by post-hoc analysis[18]. Categorical variables were analyzed using Chi-square analysis. Continuous variables were analyzed using Mann-Whitney *U* test, Friedman test for overall group significance or the independent sample t-test, depending on the distribution of data. In case of outcome variables which changed in the same direction in both the interventional and control groups, between-group comparisons were also made to determine if the change in one group was significantly greater than that of the other group.

The degree of EIB was quantified by the percentage of highest fall in FEV_1 reached after exercise during post-exercise measurements (3, 8, 15 and 30 min) as compared to the pre-exercise value. There was very high inter-individual variability in degree of EIB. Therefore the subjects were divided into exercise-sensitive (ES) and exercise-resistant (ER). Those who had a reduction of 15% or more in FEV_1 post exercise at 0-wk were considered ES, and the rest ER.

The ECP values were log-transformed prior to analysis due to their wide range. Urinary 11β-PGF2α concentrations were not normally distributed and therefore expressed in median and inter-quartile ranges. The percentage change in urinary 11β-PGF2α following exercise was used for statistical analysis.

A change of ≥0.5 in the AQOL score has been considered the Minimal Important Difference (MID) as clinically meaningful difference[19]. Based on the MID, number needed to treat (NNT) was calculated using the method described by Gyuatt et al., (1998)[20] for trials with a parallel group design.

Differences were considered significant if $p < 0.05$. All statistical calculations were performed using the SPSS software 10.0.1 release for Windows (SPSS Inc, Chicago IL, USA).

Ethical Considerations

The protocol of the study was approved by the Ethics Committee of All India Institute of Medical Sciences. The subjects signed an informed consent form before being enrolled for the study and knew that they were free to withdraw from the study at any stage without assigning any reason.

RESULTS

The demographics and baseline values of outcome variables of subjects are given in Tables 1 and 2. As seen in these tables, the two groups are comparable at baseline in every respect except PEFR expressed as percentage of the predicted value.

Table 1. Demographics and Baseline Values of Outcome Variables

	Yoga group (n = 29)	**Control group (n = 28)**	***p*-value**
Age, years (mean ± SD)	33.5 ± 11.4	33.4 ± 11.5	0.96
Males, n (%)	13 (45)	20 (71)	0.29
Females, n (%)	16 (55)	8 (29)	0.19
Weight, kg (mean ± SD)	61.3 ± 11.5	58.9 ± 9.7	0.93
Height, cm (mean ± SD)	162 ± 10.1	161.8 ± 7.4	0.40
Body Mass Index (BMI) (mean ± SD)	23.4 ± 4.3	22.6 ± 4.0	0.44
Family history of asthma present, n (%)	13 (44.8)	11 (39.3)	0.67
Duration of Asthma, years (mean ± SD)	11.6 ± 9.5	10.5 ± 11.9	0.22
Mild asthma, n (%)	13 (44.8)	11 (39.3)	0.78
Moderate asthma, n (%)	16 (55)	17 (60.7)	0.82
FVC (% of predicted)	78.7 ± 13.3	75.2 ± 15.0	0.35
FEV1 (% of predicted)	70.2 ± 17.4	62.5 ± 19.2	0.11
FEV1/FVC (% of predicted)	80.4 ± 11.5	73.7 ± 14.8	0.06
PEFR (% of predicted)	68.6 ± 18.4	57.4 ± 19.7	**0.03**
FEF25-75 (% of predicted)	38.4 ± 14.5	33.9 ± 18.3	0.31
Serum ECP level (ng/mL)	1.32 ± 0.51	1.25 ± 0.43	0.89
EIB (%fall in FEV_1 with EC)	15.0 ± 14.0	9.1 ± 10.8	0.16
Urinary 11β-PGF2α before exercise (pg/mg of creatinine), median (IQR)	455.4 (107.2–1098.1)	223.7 (98.8–562.7)	0.12

Δ Urinary 11β-PGF2α (pg/mg of creatinine) with exercise challenge, median (IQR)	26.9 (7.4–218)	38.4 (3.1–108)	0.55
AQOL total score	3.72 ± 1.17	3.64 ± 1.14	0.80
AQOL symptoms score	3.77 ± 1.34	3.62 ± 1.42	0.70
AQOL activity limitation score	3.66 ± 1.13	3.67 ± 1.17	0.95
AQOL emotional function score	3.94 ± 1.47	3.59 ± 1.39	0.35
AQOL environmental stimuli score	3.72 ± 1.17	4.04 ± 1.42	0.14

Abbreviations: AQOL, Asthma Quality of Life; FEV_1, forced expiratory volume in 1 second; $FEF_{25-75}\%$, forced mid-expiratory flow between 25% and 75% of FVC; FVC, forced vital capacity; PEFR, peak expiratory flow rate; ECP, eosinophilic cationic protein; IQR, inter- quartile range; EC, exercise challenge; EIB, exercise-induced bronchoconstriction; 11β-PGF2α, 11β prostaglandin F2α.

p-value obtained by from Chi-square analysis or 't' test or Mann-Whitney *U* test depending on the type of variables.

All values are expressed as mean ± SD, except ECP expressed as log-transformed values and 11β-PGF2α as median with inter-quartile range.

Table 2. Daily dose of inhaled corticosteroids and rescue medication at baseline

Type of medication	**Yoga (No. of subjects)**	**Control (No. of subjects)**	***p* -value**
Mean inhaled steroid dose, μg/day (mean ± SD)§	320.4 ± 158.8 (n = 17)	360 ± 151.4 (n = 15)	0.48
Rescue medication use (puffs/day)¥	2.27 ± 1.5 (n = 26)	1.98 ± 2.09 (n = 20)	0.19

§ Daily dose of inhaled steroids in beclomethasone μg equivalents over the past 1-2 wk.

These were calculated on the assumption that beclomethasone 2000 μg = budesonide 1600 μg = fluticasone 1000 μg.

¥ Rescue medication = Inhaled rescue bronchodilator intake (puffs/day) use per day and an average for 2 weeks preceding the study visit.

p-value obtained from Chi-square analysis or Mann-Whitney *U* test.

The values of the outcome measures at different time points during the study are given in Tables 3, 4, 5 and 6 and Figures 1, 2 and 3.

Table 3. Results of pulmonary function indices during the study

Parameter	***Group***	**Week 0**	**Week 2**	**Week 4**	**Week 8**	***p*** **-value**
PEFR	Yoga	68.6 ± 18.4	76.5 ± 20.5*	81.5 ± 20.9‡	85.3 ± 20.7‡	0.000
	Control	57.4 ± 19.7	58.2 ± 22.0	56.9 ± 26.2	56.2 ± 22.0	
FEV1	Yoga	70.2 ± 17.4	73.9 ± 19.6	76.1 ± 20.1	77.9 ± 17.2*	0.009
	Control	62.5 ± 19.2	63.1 ± 20.5	60.5 ± 21.6	59.9 ± 19.1	
FVC	Yoga	78.7 ± 13.4	78.0 ± 12.6	79.3 ± 13.0	82.2 ± 10.7	NS
	Control	75.0 ± 15.0	74.7 ± 18.3	73.4 ± 18.1	72.5 ± 17.5	
FEV1/ FVC	Yoga	80.4 ± 11.5	82.6 ± 13.3	83.7 ± 13.4	83.1 ± 12.2	0.011
	Control	73.7 ± 14.9	74.8 ± 13.9	72.4 ± 15.1	73.3 ± 13.8	
$FEF_{25-75\%}$	Yoga	38.4 ± 14.6	42.0 ± 19.4	45.0 ± 20.5	45.0 ± 19.7	0.035
	Control	34.0 ± 18.3	33.8 ± 17.8	32.4 ± 18.1	31.1 ± 17.1	

All values are % of predicted and expressed as mean ± SD.

* $p < .05$; † $p < .01$; ‡ $p < .001$ based on post-hoc pair wise comparisons with 0 wk values.

p-values were obtained from the between groups comparison in the GLM, NS – not significant

Table 4. Exercise induced bronchoconstriction during the study

Category	*Group*	Week 0	Week 2	Week 4	Week 8	*p* -value
% fall in FEV1 with exercise challenge	Yoga (n = 29)	15.05 ± 14.0	8.81 ± 12.9	5.65 ± 12.9	5.18 ± 11.8	0.005
	Control (n = 28)	9.13 ± 10.8	9.50 ± 11.1	10.28 ± 10.9	5.81 ± 10.1	NS
ES subjects (≥ 15% fall in FEV1)	Yoga (n = 13)	26.71 ± 12.7	13.33 ± 16.1	9.90 ± 14.5	7.95 ± 15.1	0.003
	Control (n = 9)	21.47 ± 6.4	8.83 ± 9.8	18.69* ± 8.2	12.71 ± 9.1	0.015
ER subjects (<15% fall in FEV1)	Yoga (n = 16)	5.58 ± 5.1	5.13 ± 8.4	2.20 ± 10.6	2.93 ± 8.0	NS
	Control (n = 19)	3.29 ± 6.6	9.82 ± 11.9	6.31 ± 9.8	2.53 ± 9.0	NS

All values are expressed as mean ± SD. ES – exercise sensitive; ER – exercise resistant.

P-value was based on Friedman test applied on % fall in FEV1 after exercise challenge within each group separately.

* Between groups comparisons were made by Mann-Whitney *U* test at individual visits, there was significant difference observed at 4 wk between groups in exercise-sensitive subjects (p = 0.025).

Table 5. Normalized values of Δ urinary 11β-PGF2α with exercise challenge

Category	*Group*	Week 0	Week 2	Week 4	Week 8	*p* - value [a]
% change in 11β-PGF2α excretion. (All subjects)	Yoga (n = 29)	17.6 ± 18.6	19.7 ± 24.7	14.7 ± 16.5	16.8 ± 19.7	0.49
	Control (n = 28)	16.6 ± 21.0	18.6 ± 22.0	16.3 ± 22.3	20.7 ± 23.5	0.84
ES subjects (mean ± SD)	Yoga (n = 13)	18.3 ± 22.2	20.4 ± 29.5	13.8 ± 16.8	12.3 ± 16.9	0.63

	Control (n = 9)	15.5 ± 21.2	13.3 ± 18.1	28.5 ± 16.3*	18.9 ± 8.2	0.11
ER subjects (mean ± SD)	Yoga (n = 16)	17.1 ± 15.9	19.0 ± 21.0	15.4 ± 16.8	20.4 ± 21.5	0.79
	Control (n = 19)	17.0 ± 21.5	21.1 ± 23.6	10.6 ± 22.8	21.6 ± 28.2	0.08

All values (normalized) are expressed as % change in 11β prostaglandin F2α excretion with exercise challenge (mean ± SD). ES – exercise sensitive; ER – exercise resistant subjects.

[a] *P*-value was based on Friedman test in each group separately from % change in 11β-PGF2α with exercise.

*p < .05 based on Mann-Whitney *U* test at 4 wk between two groups.

Table 6. Asthma quality of life scores during the study

AQOL sub domains	***Group***	**Week 0**	**Week 2**	**Week 4**	**Week 8**	***p*** **-value**
Symptoms	Yoga	3.77 ± 1.3	5.07 ± 1.4‡	5.38 ± 1.1‡	5.42 ± 1.2‡	0.033
	Control	3.63 ± 1.4	3.96 ± 1.7	4.42 ± 1.5†	4.70 ± 1.7†	
Activity limitation	Yoga	3.66 ± 1.1	4.82 ± 1.3‡	5.21 ± 1.0‡	5.47 ± 1.1‡	0.033
	Control	3.67 ± 1.2	3.74 ± 1.5	3.90 ± 1.4	4.20 ± 1.5	
Emotional function	Yoga	3.94 ± 1.5	5.10 ± 1.6†	5.45 ± 1.3‡	5.71 ± 1.3‡	0.006
	Control	3.60 ± 1.4	4.01 ± 1.6	4.15 ± 1.8	4.32 ± 1.7	
Response to environmental stimuli	Yoga	3.46 ± 1.5	4.64 ± 1.5‡	5.0 ± 1.4‡	5.3 ± 1.6‡	NS
	Control	4.04 ± 1.4	3.83 ± 1.5	4.08 ± 1.6	4.40 ± 1.8	
Total Quality of Life	Yoga	3.72 ± 1.2	4.93 ± 1.3‡	5.28 ± 1.0‡	5.46 ± 1.1‡	0.013

	Control	3.64 ± 1.1	3.90 ± 1.5	4.17 ± 1.4*	4.50 ± 1.5†	

All values are expressed as mean ± SD.

* p < .05; † p < .01; ‡ p < .001 based on post-hoc pair wise comparisons with 0 wk values.

p-values were obtained from the between groups comparison in the GLM, NS – not significant.

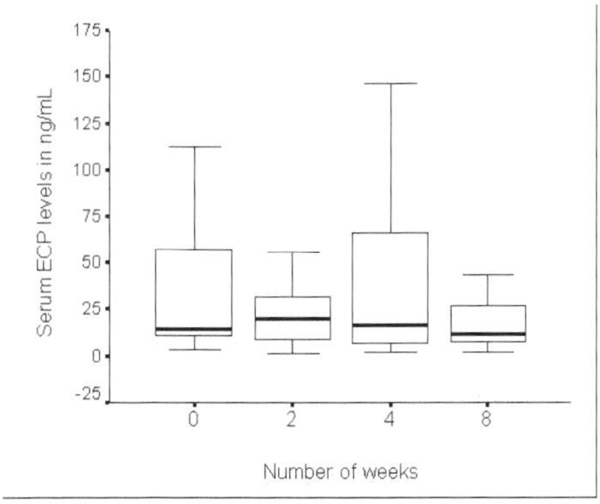

Figure 1. Serum eosinophilic cationic protein in yoga group during the study. Serum eosinophilic cationic protein in yoga group (n = 29) during 8 wk study period. Horizontal thick bars indicate median values and box the inter-quartile range (IQR).

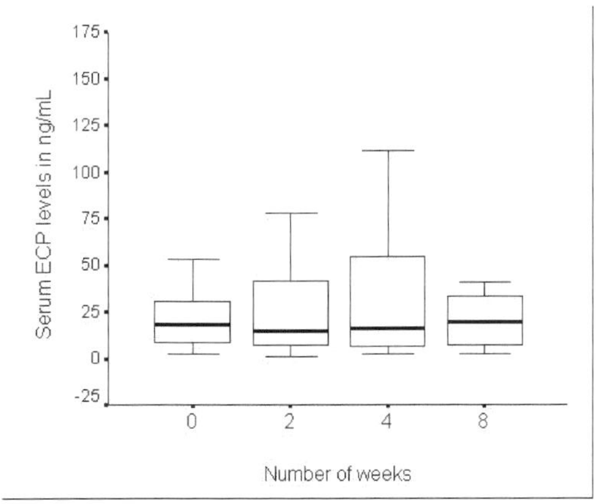

Figure 2. Serum eosinophilic cationic protein in control group during the study. Serum eosinophilic cationic protein in control group (n = 28) during 8 wk study period

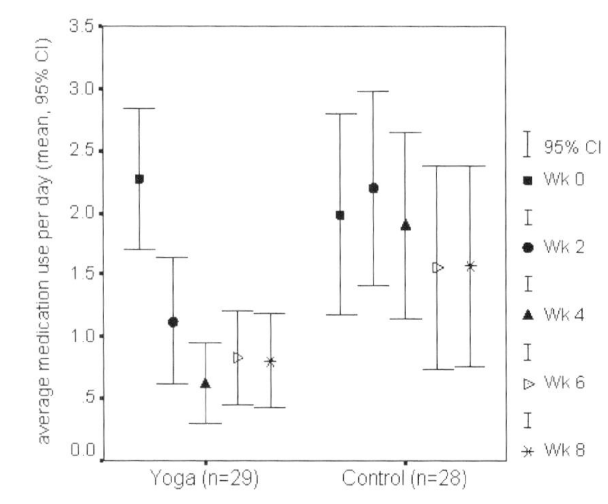

Figure 3. Frequency of rescue medication use in yoga and control groups. Rescue medication use was calculated as an average use of short acting β2-agonists per day (puffs/day) for 2 wk, hence, there are five time points. There is significant overall decline in rescue medication use in both yoga ($p < .001$) and control groups ($p < .05$). Also significant differences seen at 2 wk ($p < .05$) and 4 wk ($p < .01$) between yoga and control groups.

There were no significant changes in pulmonary function indices in the control group during the course of the study. In contrast, there was a steady and progressive improvement in the subjects of yoga group during the same period. The change in the yoga group was statistically significant for FEV_1 at 8 wk and PEFR at 2, 4 and 8 wk, as compared to the corresponding baseline values (Table 3). In addition, significant group mean differences were observed between yoga and control groups over the 8-wk study period in PEFR, FEV_1, FEV_1/FVC and $FEF_{25-75\%}$. In PEFR, since there is a significant baseline difference, baseline was considered as a constant covariate for between-group comparisons.

There was no significant change in serum ECP levels during the 8-wk study period in either group (Figures 1 and 2). The EIB decreased significantly in the yoga group, but not in the control group over the 8-wk study period (Table 4). Subgroup analysis in ES subjects revealed significant reduction in EIB over the 8-wk study period in both yoga and control groups but the amount of reduction observed was more marked in the yoga group than in the control group. In ER subjects the change in EIB was not significant in either group. Similar analysis of percentage change in urinary 11β-PGF2α levels in response to the exercise challenge revealed no significant change during the study in either group (Table 5). One interesting observation at 4-wk in the control group (in ES subjects) was that the EIB was rather significantly high, and this was associated with a significant increase in urinary 11β-PGF2α excretion in response to the exercise challenge as compared to the ES subjects in yoga group (Tables 4 and 5).

The AQOL scores showed an improvement over the 8-wk study period in both groups (Table 6). But the improvement was achieved relatively earlier and was more complete in the yoga group. At 2 wk, the improvement was statistically significant as compared to the baseline score in the total quality of life as well as each of the four sub-domains in the yoga group, but not in the control group. Even by 8-wk, significant improvement was seen in the control group only in total AQOL and symptoms but not in activity limitation, emotional function or the response to environmental stimuli. The NNT was found to be 1.82 for the total score, 2.41 for symptoms score, 1.66 for activity limitation score, 1.91 for emotional function score and 1.70 for the response to environmental stimuli score, which means ≤ 2 subjects needed to be treated with yoga plus conventional treatment for one subject to have a clinically meaningful improvement in quality of life over and above the improvement that the subject would have experienced with conventional treatment alone.

The frequency of rescue meditation use showed a significant decrease over the 8-wk study period in both the groups. But the decrease in rescue meditation use was achieved relatively earlier and was more marked in the yoga group than in the control group (Figure 3). Rescue medication use between yoga and control groups was significantly different at 2 and 4 wk because of the remarkable decrease in the frequency of the medication use only in the yoga group.

Although the level of compliance for yoga practice aimed at was at least 5 days a week, this level could not be sustained throughout the 8 weeks of the study. The compliance for yoga practice as reported by the subject was (average number of practice days per week): Wk 0, 4.6 ± 0.5; Wk 2, 4.6 ± 1.1; Wk 4, 4.5 ± 1.2; and Wk 8, 4.4 ± 1.1.

DISCUSSION

The present study has demonstrated that adding the mind body approach of yoga to the predominantly physical approach of conventional care results in measurable improvement in some subjective as well as objective outcomes in bronchial asthma. There have been previous RCTs on the efficacy of yoga which have arrived at similar conclusions[3, 6, 10]. But none of the previous trials have gone into the plausible mechanisms by which yoga might help bronchial asthma even in a small way as the present study has done. Since the major built-in difference between the two groups was the yogic intervention during 8-wk of the study, any difference in the outcome is likely to be due to yoga. The control being a wait-listed group, it does not control completely for non-specific effects of the experimental intervention such as expectation, belief, subject investment in time and effort, change in the subjects' daily routine, investigator interaction, social interaction and support, and placebo effect.

The 8-wk duration of the yogic intervention was an arbitrary compromise. It is long enough to demonstrate the effect, but the ideal would be a much longer study. However, in practice the decline in compliance increases steeply as the study gets longer, possibly owing to the time commitment required from the subjects[9]. Since there is no appropriate placebo possible for yoga, the subject cannot be blinded. The allocation of the subjects to the two groups was randomized, but the selection of subjects from the patient population was not random. Every subject who made it to our clinic to volunteer for the trial was already convinced in his mind about the beneficial effects of yoga because of its popularity in India. This is an inherent flaw of such trials, and limits the extrapolation of results to the entire subject population. However, this limitation does not reduce the external validity of the study because in practice only those subjects opt for a particular modality of treatment who have faith in its efficacy.

We chose, keeping the subjects' convenience and our infrastructure in mind, a 2-wk supervised out-patient intervention at IHC followed by 6 wk of closely monitored intervention at home. The 2-wk training period was adequate to make the subjects reasonably proficient in physical postures, breathing exercises, meditation and relaxation techniques, and to acquaint them with principles of stress management and a healthy yoga-based lifestyle. Leaving the subject to continue with these practices on his own and apply the principles learnt to everyday situations for 6 wk was reasonably long to test the feasibility of the intervention in real life, and to assess how far the gains achieved in 2 wk of supervised training were sustained during follow-up. Not only could the subjects continue with the intervention, the gains of the first 2-wk were improved upon during the 6-wk follow up.

The various interventions administered under the omnibus title 'yoga' cover a very wide spectrum. This is inevitable because yoga is not a system of medicine; it is a way of life, the implications of which go beyond health and disease[21]. When yoga is employed for prevention or management of disease, some convenient and highly visible elements of yoga are used selectively. Our intervention has used yoga as a tool in mind-body medicine[22]. This approach now has a strong scientific foundation in psychoneuroimmunology, and is particularly relevant to a disorder like bronchial asthma which is characterized by deranged immune function.

The results of the present study allow a few fairly firm conclusions. Significant, steady and progressive improvement in key objective variables such as PEFR and FEV_1 only in the yoga group but not in the control group indicates the efficacy of yoga. This is further substantiated by the significantly greater improvement in quality of life in the yoga group than in the control group.

In yoga group, there was also a significant trend towards improvement in exercise tolerance assessed by the exercise-induced fall in FEV_1, which became more evident in ES subjects. The effects of yoga on EIB in our study are similar to those reported by Jain *et al*. (1991)[5], who demonstrated a significant increase in exercise tolerance as indicated by a decrease in exercise-induced bronchial lability index (BLI) with yoga practice for 40-days in adolescents with mild to severe asthma and in another study by the same authors in moderate to severe chronic asthma patients[23].

Since the blood level of ECP is directly related to the severity of asthma and inversely related to lung functions, we have used ECP as a peripheral marker in understanding the disease progress [24, 25]. In our present study, no significant changes were observed in ECP in either group. The high variability in ECP levels suggests the need for a much larger number of subjects for arriving at definitive conclusions.

Apart from the IgE mediated mast cell activation[26], mast cells may also be activated independently of high affinity IgE receptor (FcεRI), by emotional stress and it can induce asthma exacerbations [27, 28]. The possibility is made more likely because psychological or emotional stress triggers dura mast cell secretion via release of corticotropin-releasing hormone, CRH[29]. Another study also indicated that acute psychological stress could lead to cardiac mast cell degranulation through CRH acting directly or through neurotensin[30]. Therefore, we speculated that yoga might play a vital role in reducing the mast cell activation levels by improving the emotional function of asthmatics thereby reducing the inflammatory condition and would cause the favorable effects on this outcome measure.

Another plausible explanation for the role of yoga in reducing the mast cell degranulation could be based on the frictional stress from air flowing through narrowed airways damaging the airway mucosa and thereby perpetuating airway inflammation and airway obstruction. At high air flow rates, high values of the frictional stress could damage the airway wall, especially during episodes of cough, and particularly when the mucosa is inflamed and friable as it is in asthmatic patients[31]. The slow and gentle breathing in some of the pranayamas (breathing practices in yoga) may reverse the process[32] by reducing the frictional stress, and thereby stabilizing the mast cell degranulation. Olness *et al*. (1999)[33] also demonstrated an inhibition of mast cell activation by conditioned relaxation techniques such as relaxation-imagery in children having migraine. In the present study, the urinary concentrations of 11β-PGF2α against exercise stimuli in yoga and control groups did not show any significant trend towards the inhibition of mast cell activation except a significant difference observed between groups at 4 wk in ES subjects (Table 5). Therefore, our study could not examine conclusively the above speculations because of high variability of 11β-PGF2α and a small sample size (yoga, 13; control, 9) in ES subjects. In addition, the exercise in the present study was a much less vigorous stimulus for provocation of airways and activating the mast cells than allergen provocation in which approximately 200–300% increase is seen in levels of urinary 11β-PGF2α with an allergen challenge[34].

The favorable effect of yoga in bronchial asthma observed by us in some outcome measures is consistent with similar results reported in several previous studies[3–5, 10]. Most of these, except the study by Jain *et al* (1991)[5], were also randomized controlled trials. But lately there have been a few studies which have reported either negative or very weak effects of yoga[6–9, 35]. All these were also controlled trials, and the study designs are apparently more rigorous than in the earlier trials. However, that does not necessarily mean that the positive results of earlier studies were due to insufficient rigor. Rather, it seems that in order to meet the exacting demands of methodology, some of the recent studies have overlooked some basic features of yoga. Moreover, yoga is much more than breathing exercises. It is only an integrated practice e.g. that given in the study by Nagarathna and Nagendra (1985)[10] or in the present study, which comes closest in an experimental situation to what yoga really is. In the study by

Sabina *et al* (2005)[9], yoga has been equated with postures or breath work. The frequency is only twice a week without any serious monitoring of the frequency with which the practices are repeated at home on other days. Moreover, the control is a stretch. If yogic stretches are also treated like just a stretch, a non-yogic stretch could be equally good. It is only when the yogic stretch is performed with a yogic attitude, and preferably, when it becomes a component in the yogic way and view of life, that the yogic stretch acquires superiority over a non-yogic stretch[36]. That is possibly the reason why in the study by Sabina *et al* (2005)[9], both groups improved but the difference between the groups was not significant. In the most recent studies[8, 9], the scientific rigor has been improved upon by giving to the control group an intervention similar to that given to the yoga group, and the trials have been masked. Masking in a study which involves highly visible physical practices is meaningless. Further, if the interventions in the two groups are very similar, the outcomes in the two groups cannot be very different. The mere fact that in one group the techniques have been selected from a yogic text and in the other from a text on physiotherapy or modern psychology will not affect the outcome. Further, the comprehensive approach to yoga has been avoided probably to keep the intervention well-defined. But in the process of testing a holistic treatment using a design which is ideal for testing a drug, the holistic treatment should not be reduced to one of its components. For the sake of fitting our design into a rigid framework, we should not end up evaluating a modality of treatment which is different from the one we have set out to examine. That answers also one of the commonest criticisms leveraged at studies like the present one i.e. the study does not allow us to determine the proportionate contribution of each component of the intervention. To view a holistic treatment as a combination of components is convenient but inappropriate. The efficacy observed is that of the whole intervention, and may be greater than the sum of the efficacy of its components.

The present study is the first one to attempt an investigation into the mechanism by which yoga helps bronchial asthma. Although it has not succeeded in throwing much light in that direction, there are several other subtle reasons why this study and many more similar studies in the future would be justified. Although there are several studies available on the efficacy of yoga in bronchial asthma and several other diseases, studies have to reach a critical mass before they can influence practice. Further, in case of yoga, very few of the studies are randomized controlled trials[1]. Because of the dilution of a few RCTs by several less rigorous trials, and also because of widespread scepticism with which disciplines like yoga are approached by practitioners of conventional medicine, the impact of the RCTs is inadequate. Further, it would help to integrate mind-body approaches like yoga into the practice of scientific medicine if the trials not only show that these approaches work, but also how they work in terms of measurable basic mechanisms.

CONCLUSION

The present RCT has shown that adding a comprehensive yoga-based mind-body intervention to the conventional treatment improves several measures of pulmonary function in subjects having mild to moderate bronchial asthma. There is a significant trend suggesting a decrease in exercise-induced bronchoconstriction in the yoga group, particularly in the exercise-sensitive subjects. Yoga and control groups did not show a clear trend towards the inhibition of mast cell activation except for a significant difference observed between groups at 4 wk in exercise-sensitive subjects. Once again, in view of the high variability in response, a much larger number of exercise-sensitive subjects are required for definitive conclusions. Yoga improves the quality of life and reduces rescue medication use in bronchial asthma, and achieves the reduction earlier than conventional treatment alone. The trial supports the use of yoga in the management of bronchial asthma.

REFERENCES

1. Khalsa SB: Yoga as a therapeutic intervention: a bibliometric analysis of published research studies. Indian J Physiol Pharmacol. 2004, 48: 269-285.
2. Bhole MV: Treatment of Bronchial Asthma by Yogic Methods. Yoga Mimansa. 1967, 9: 33-41.
3. WilsonAF, HonsbergerR, ChiuJT, NoveyHS: Transcendental meditation and asthma. Respiration. 1975, 32: 74-80. 10.1159/000193637.
4. Singh V, Wisniewski A, Britton J, Tattersfield A: Effect of yoga breathing exercises (pranayama) on airway reactivity in subjects with asthma. Lancet. 1990, 335: 1381-1383. 10.1016/0140-6736(90)91254-8.
5. Jain SC, Rai L, Valecha A, Jha UK, Bhatnagar SO, Ram K: Effect of yoga training on exercise tolerance in adolescents with childhood asthma. J Asthma. 1991, 28: 437-442. 10.3109/02770909109110627.
6. Vedanthan PK, Kesavalu LN, Murthy KC, Duvall K, Hall MJ, Baker S, Nagarathna S: Clinical study of yoga techniques in university students with asthma: a controlled study. Allergy Asthma Proc. 1998, 19: 3-9. 10.2500/108854198778557971.
7. Cooper S, Oborne J, Newton S, Harrison V, Thompson Coon J, Lewis S, Tattersfield A: Effect of two breathing exercises (Buteyko and pranayama) in asthma: a randomised controlled trial. Thorax. 2003, 58: 674-679. 10.1136/thorax.58.8.674.
8. Manocha R, Marks GB, Kenchington P, Peters D, Salome CM: Sahaja yoga in the management of moderate to severe asthma: a randomised controlled trial. Thorax. 2002, 57: 110-115. 10.1136/thorax.57.2.110.
9. Sabina AB, Williams AL, Wall HK, Bansal S, Chupp G, Katz DL: Yoga intervention for adults with mild-to-moderate asthma: a pilot study. Ann Allergy Asthma Immunol. 2005, 94: 543-548.
10. Nagarathna R, Nagendra HR: Yoga for bronchial asthma: a controlled study. Br Med J (Clin Res Ed). 1985, 291: 1077-1079. 10.1136/bmj.291.6502.1077.
11. American Thoracic Society: Standardization of spirometry, 1994 update. Am J Respir Crit Care Med. 1995, 152: 1107-1136.
12. Ferguson AC, Vaughan R, Brown H, Curtis C: Evaluation of serum eosinophilic cationic protein as a marker of disease activity in chronic asthma. J Allergy Clin Immunol. 1995, 95: 23-28. 10.1016/S0091-6749(95)70148-6.

13. Flores O, Sun L, Vaziri N, Miyada D: Colorimetric rate method for the determination of creatinine as implemented by the Beckman Creatinine Analyzer 2. Am J Med Technol. 1980, 46: 792-798.

14. Misso N, Aggarwal S, Phelps S, Beard R, Thompson PJ: Urinary leukotriene E4 and 9 alpha, 11 beta-prostaglandin F2 concentrations in mild, moderate and severe asthma, and in healthy subjects. Clin Exp Allergy. 2004, 34: 624-631. 10.1111/j.1365-2222.2004.1912.x.

15. Juniper EF, Gyuatt GH, Epstein RS, Ferrie PJ, Jaeschke R, Hiller TK: Evaluation of impairment of health related quality of life in asthma: development of a questionnaire for use in clinical trials. Thorax. 1992, 47: 76-83. 10.1136/thx.47.2.76.

16. Juniper E, Gyuatt G, Ferrie PJ, Griffith LE: Measuring quality of life in asthma. Am Rev Respir Dis. 1993, 147: 832-838.

17. Chhabra S, Kaushik S: Validation of the asthma quality of life questionnaire (AQLQ-UK English version) in Indian asthmatic subjects. Indian J Chest Dis Allied Sci. 2005, 47: 167-173.

18. Bonferroni CE: Il calcolo delle assicurazioni su gruppi di teste. Studi in Onore del Professore Salvatore Ortu Carboni: Rome: Italy. 1935, 13-60.

19. Juniper E, Guyatt G, Willan A, Griffith LE: Determining a minimal important change in a disease-specific quality of life questionnaire. J Clin Epidemiol. 1994, 47: 81-87. 10.1016/0895-4356(94)90036-1.

20. Gyuatt G, Juniper E, Walter S, Griffith L, Goldstein R: Interpreting treatment effects in randomised trials. Br Med J. 1998, 101: 163-170.

21. Bijlani RL: Is yoga system of medicine?. New Approaches to Medicine and Health. 1998, 6: 27-36.

22. Bijlani RL: Body-mind medicine in action: the Integral Health Clinic. New Approaches to Medicine and Health. 2004, 12: 37-48.

23. Jain SC, Talukdar B: Evaluation of yoga therapy programme for patients of bronchial asthma. Singapore Med J. 1993, 34: 306-308.

24. Motojima S, Tateishi K, Koseki T, Makino S, Fukuda T: Serum levels of eosinophil cationic protein and IL-5 in patients with asthma without systemic corticosteroids. Int Arch Allergy Immunol. 1997, 114: 55-59.

25. Wever A, Hess J, Hermans J: The use of serum eosinophil cationic protein (ECP) in the management of steroid therapy in chronic asthma. Clin Exp Allergy. 1997, 27: 519-529. 10.1111/j.1365-2222.1997.tb00740.x.

26. Ishizaka T, Sian C, Ishizaka K: Complement fixation by aggregated IgE through alternate pathway. J Immunol. 1972, 108: 848-851.

27. Bienenstock J: Stress and asthma: the plot thickens. Am J Respir Crit Care Med. 2002, 165: 1034-1035.

28. Bienenstock J, MacQueen G, Sestini P, Marshall JS, Stead RH, Perdue MH: Mast cell/nerve interactions in vitro and in vivo. Am Rev Respir Dis. 1991, 143: S55-58.

29. Theoharides TC, Spanos C, Pang X, Alferes L, Ligris K, Letourneau R, Rozniecki JJ, Webster E, Chrousos GP: Stress-induced intracranial mast cell degranulation: a corticotropin-releasing hormone-mediated effect. Endocrinology. 1995, 136: 5745-5750. 10.1210/en.136.12.5745.

30. Pang X, Alexacos N, Letourneau R, Seretakis D, Gao W, Boucher W, Cochrane DE, Theoharides TC: A neurotensin receptor antagonist inhibits acute immobilization stress-induced cardiac mast cell degranulation, a corticotropin-releasing hormone-dependent process. J Pharmacol Exp Ther. 1998, 287: 307-314.

31. Chowdhary R, Singh V, Tattersfield AE, Sharma SD, Kar S, Gupta AB: Relationship of flow and cross-sectional area to frictional stress in airway models of asthma. J Asthma. 1999, 36: R419-426. 10.3109/02770909909087284.

32. Singh V, Chowdhary R, Chowdhary N: The role of cough and hyperventilation in perpetuating airway inflammation in asthma. J Assoc Physicians India. 2000, 48: 343-345.

33. Olness K, Hall H, Rozniecki JJ, Schmidt W, Theoharides TC: Mast Cell Activation in Children with Migraine Before and After Training in Self-regulation. Headache. 1999, 39: 101-107. 10.1046/j.1526-4610.1999.3902101.x.

34. O'Sullivan S, Dahlen B, Dahlen SE, Kumlin M: Increased urinary excretion of the prostaglandin D2 metabolite 9 alpha, 11 beta-prostaglandin F2 after aspirin challenge supports mast cell activation in aspirin-induced airway obstruction. J Allergy Clin Immunol. 1996, 98: 421-432. 10.1016/S0091-6749(96)70167-7.

35. Khanam AA, Sachdeva U, Guleria R, Deepak KK: Study of pulmonary and autonomic functions of asthma patients after yoga training. Indian J Physiol Pharmacol. 1996, 40: 318-324.

Chapter 8

12 Weeks of Kindergarten-Based Yoga Practice Increases Visual Attention, Visual-Motor Precision and Decreases Behavior of Inattention and Hyperactivity in 5-Year-Old Children

Sana Jarraya[1], Matthias Wagner[2] , Mohamed Jarraya[1] and Florian A. Engel[3]

[1] Research Unit, High Institute of Sport and Physical Education, University of Sfax, Sfax, Tunisia

[2] Department of Sport Science, Bundeswehr University Munich, Neubiberg, Germany

[3] Department Movement and Training Science, Institute of Sport and Sport Science, Heidelberg University, Heidelberg, Germany

Citation: 12 Weeks of Kindergarten-Based Yoga Practice Increases Visual Attention, Visual-Motor Precision and Decreases Behavior of Inattention and Hyperactivity in 5-Year-Old Children. Jarraya S, Wagner M, Jarraya M and Engel FA (2019). Front. Psychol. 10:796. doi: 10.3389/fpsyg.2019.00796.

ABSTRACT

The present study assesses the impact of Kindergarten-based yoga on cognitive performance, visual-motor coordination, and behavior of inattention and hyperactivity in 5-year-old children. In this randomized controlled trial, 45 children (28 female; 17 male; 5.2 ± 0.4 years) participated. Over 12 weeks, 15 children performed Hatha-yoga twice a week for 30 min, another 15 children performed generic physical education (PE) twice a week for 30 min, and 15 children performed no kind of physical activities, serving as control group (CG). Prior to (T_0) and after 12 weeks (T_1), all participants completed Visual Attention and Visuomotor Precision subtests of Neuropsychological Evaluation Battery and teachers evaluated children's behavior of inattention and hyperactivity with the Attention-Deficit/Hyperactivity Disorder (ADHD) Rating Scale-IV. At T_0, no significant differences between groups appeared. Repeated measures analysis of variance revealed that following Bonferroni–Holm corrections yoga, in comparison to PE and CG, had a significant positive impact on the development on behavior of inattention and hyperactivity. Further, yoga has a significant positive impact on completion times in two visumotor precision tasks in comparison to PE. Finally, results indicate a significant positive effect of yoga on visual attention scores in comparison to CG. 12 weeks of Kindergarten-based yoga improves selected visual attention and visual-motor precision parameters and decreases behavior of inattention and hyperactivity in 5-year-old children. Consequently, yoga represents a sufficient and cost-benefit effective exercise which could enhance cognitive and behavioral factors relevant for learning and academic achievement among young children.

INTRODUCTION

As an important basis for academic achievement in children and adolescents, adequate cognitive functions are suggested (Duncan et al., 2007; McClelland et al., 2013). Whereas, cognitive functions are defined as an individual's mental processes, including memory, attention, learning, problem solving, reasoning, and decision-making (Matlin, 2009).

Recently, a growing body of literature appears examining the effects of acute and chronic physical activity in young and preadolescent children and its impacts on cognitive and executive functions as well as on academic achievement (Verburgh et al., 2014; Ludyga et al., 2016; de Greeff et al., 2018; Singh et al., 2018). Whereas, the association between executive functions and academic achievement is consistently demonstrated and emphasized (Blair and Raver, 2015). The concept of executive functions comprises several cognitive functions and refers to at least three interrelated cognitive functions (Miyake et al., 2000): inhibition, shifting, and updating (for overview, see Diamond, 2013). In this context, recent research demonstrated that physical activity can lead to improved cognitive and executive functions and enhanced academic achievement in young students (Sattelmair and Ratey, 2009; Fedewa and Ahn, 2011; Verburgh et al., 2014; de Greeff et al., 2018; Oberer et al., 2018). More precisely, a recent meta-analysis showed that physical activity has a small positive effect on school engagement in students aged 6–15 years (Owen et al., 2016). Since school engagement is a crucial factor determining academic performance (Perry et al., 2010), this demonstrates the potential of physical activity for academic achievement. Furthermore, reviewed data suggest that in young preadolescent students (6–12 years), acute physical activity has a positive effect on attention, while chronic physical activity reveals a positive effect on executive functions, attention, and academic performance (de Greeff et al., 2018). In this context, Alesi et al. (2014) demonstrated that, as a result of regular Karate practice, 9-year-old Karate athletes exhibited better working memory and visual selective attention compared to their sedentary peers.

The impact of physical exercise on deficits of attention in children is a related subject and of relevance to the present study, because disturbances revealing a deficit of attention are displayed as one of the reasons of academic failure as much as they interfere with requirements of the school (Flessas and Lussier, 2001). Rates of Attention-Deficit/Hyperactivity Disorder (ADHD) in preschool children are rising constantly and ADHD prevalence is estimated to be at 2.1% among U.S. American preschool children (Danielson et al., 2018). ADHD is a neurodevelopmental disorder characterized by symptoms of inattention and hyperactivity/impulsivity, which potentially causes lifetime impairment for afflicted individuals (American Psychiatric Association [APA], 2013). Since attention is a cognitive function and strongly implicated in the processes of learning (Flessas and Lussier, 2001) for both simple and complex activities (Flessas and Lussier, 2001; Delvolvé, 2005), we intended to analyze how a sensory-motor training in children influences behavior of inattention and hyperactivity and cognitive performance.

In this context, numerous studies demonstrated the significant effects of a sensory-motor training in children in terms of (i) decreased hyperactivity, anxiety, depression, and aggression (Banaschewski et al., 2001; Rosen et al., 2015) and (ii) reduction of behavioral and emotional problems (Sibley and Etnier, 2003; Haapala, 2012; Singh et al., 2012; Lees and Hopkins, 2013; MacCobb et al., 2014; Sandford et al., 2015). Consequently, physical activities and appropriate training programs are likely to boost cognitive functions (Hillman et al., 2014; Schmidt et al., 2015; Altenburg et al., 2016) and improve behavior and mood (Delvolvé, 2005; Gothe et al., 2013; MacCobb et al., 2014), whereby younger children seem to reveal higher benefits (Leconte, 2005).

Yoga is suggested to be an effective sensory-motor training for children to reduce behavioral and emotional problems and improve cognitive functions which might have a positive impact on academic performance (Chaya et al., 2012; Chou and Huang, 2017; Bazzano et al., 2018; Cohen et al., 2018). Recent studies demonstrated that yoga is one intervention with the potential to increase attention and academic performance in children (Chaya et al., 2012; Khalsa et al., 2012; Noggle et al., 2012; Chou and Huang, 2017) and adolescents (Khalsa et al., 2012; Noggle et al., 2012; Purohit and Pradhan, 2016). Yoga practice includes stretching postures, breathing exercises, and meditation (Taimini, 2005; Brethenoux-Seguin, 2007; Rosen et al., 2015). Indeed, conscious breathing has an impact on the parasympathetic system, causing a relaxation (Innes et al., 2005). Since yoga is a conscious exercise that builds attention control and inhibitory skills (Kaunhoven and Dorjee, 2017), it may have the potential to increase attention span on task during class and may represent an enjoyable and cost-effective treatment for children with and without attention problems (Peck et al., 2005).

Indeed, a growing body of evidence evolved underpinning the potential of yoga to support cognitive functioning in children and adolescents in different settings. Yoga was effective in improving delayed recall of spatial information and verbal memory in 11–16-year-old children (Manjunath and Telles, 2004). Similar results were reported for three months of yoga practice on executive functions of adolescent dwelling in an orphan home (Purohit and Pradhan, 2016). Yoga practice was an effective tool when working with students with Down syndrome, cerebral palsy, autism, sensory integration disorder, learning difficulties (Galantino et al., 2008), and attention deficit hyperactivity disorder (Balasubramaniam et al., 2012). In addition, yoga revealed a positive impact on cognitive functions in children with attention deficit and hyperactive disorder (American Psychiatric Association [APA], 2000; Harrison et al., 2004; Jensen and Kenny, 2004; Krisanaprakornkit et al., 2010; Diamond, 2012). Yoga is also associated with a positive effect on depression, anxiety, mod, self-esteem, and higher academic performance in children (Ortega et al., 2008). Additionally, yoga reduces negativity and aggression in children (Santangelo, 2012), improves social behavior in children aged 6–8 (Folleto et al., 2016), increases spatial memory scores, strategic planning, and concentration in children (Telles et al., 1993), as well as it improves cognitive performance of 7–9-year-old school children (Chaya et al., 2012).

Nevertheless, research involving yoga and Kindergarten children is relatively sparse. Therefore, the main purpose of the present study is to analyze the impact of yoga practice in comparison to generic physical education (PE) and a passive control group (CG) on (i) attention as a cognitive performance (operationalized with the visual attention test) and visual-motor coordination (operationalized with the Visuomotor precision test) and (ii) ADHD relevant behavior of inattention and hyperactivity (operationalized with the ADHD Rating Scale-IV), in 5-year-old preschool Tunisian children.

We hypothesized that in comparison to the generic PE group as well as the passive CG, the yoga group would display a significant improvement in visual attention and visual-motor precision as well as a significant reduction in inattention and hyperactivity behavior.

MATERIALS AND METHODS

Participants

In this single-center, three-arm randomized, controlled study, 45 healthy children (28 female; 17 male; 5.2 ± 0.4 years) of a private Tunisian Kindergarten volunteered to participate. The Kindergarten is located in an urban setting and participants were from middle class families with a corresponding average to high socio-economic status. All participants and their legal guardians were informed in detail, in written form as well as orally, about the design of the study, including the potential risks and benefits. Subsequently, parents, respectively, legal guardians of the children, provided their informed written consent to participate. Participants were free to withdraw from the study at any time without further consequences. The inclusion criteria were a lack of any frequent participation in yoga exercise programs for at least 6 months prior to the study; no daily intake of medication; and for inclusion in the analysis, completion of at least 80% of the yoga sessions.

All procedures were conducted in accordance with the Code of Ethics for human experimentation of the World Medical Association, the

Declaration of Helsinki (World Medical Association, 2013), as well as the ethical standards of the International Journal of Sports Medecine (Harriss and Atkinson, 2015). The experimental protocol was pre-approved by the ethical review board of the High Institute of Sport and Physical Education of University of Sfax, Tunisia.

Overall Study Design

Fifteen students performed a total of 24 yoga sessions, during a period of 12 weeks (two yoga sessions per week, 30 min per session) during regular Kindergarten hours, 15 students performed a generic PE program (24 sessions, two sessions per week, 30 min per session), and 15 students served as passive CG (Figure 1). Participants were randomly assigned to one of the three groups. Prior to (T_0) and after the 12 weeks of either yoga, PE, or passive control (T_1), all participants completed one subtest of the Developmental Neuropsychological Assessment (NEPSY) (Korkman et al., 1998) and one subtest of NEPSY-II (Korkman et al., 2007) and the teachers evaluated the inattention and hyperactivity behaviors of the participants with the help of the school version of the ADHD Rating Scale-IV (DuPaul et al., 1998). Since Kindergarten teachers evaluated the children's behavior with the help of the ADHD Rating Scale-IV, it was necessary to warrant impartiality of Kindergarten teachers toward the children. Therefore, Kindergarten teachers were not informed which children were assigned to the respective groups. Furthermore, teachers were not fully informed about the purpose of the study as well as about the nature and content of the two different interventions. The post-intervention testing started 72 h after the final scheduled session of yoga and PE (Figure 1).

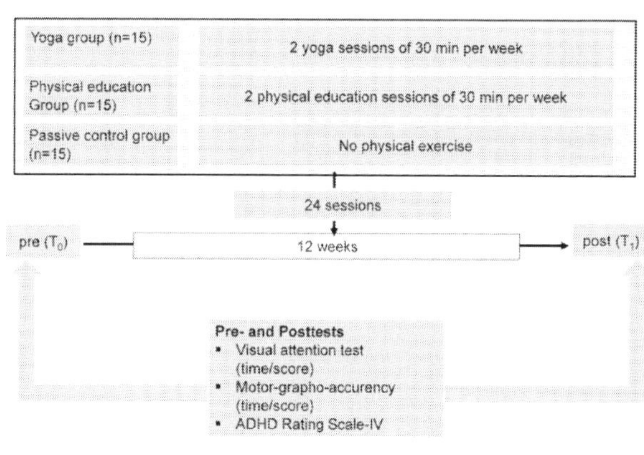

Figure 1. Overview of the study protocols with pre- and post-tests.

The Interventions

The yoga group performed 24 yoga sessions (two yoga sessions per week, 30 min per session) involving Hatha yoga. An adapted Hatha yoga program was applied to meet the level of the children and the goals of the research. The yoga program was conducted by a certified yoga teacher. Each session during the 12-week intervention involved a 30-min yoga program in the Kindergarten's gym. Each session included a 5 min preparatory period as a warm-up, comprising jogging and jumping followed by yoga specific stretching and loosening exercises and breathing exercises. During the main part of yoga sessions children performed yoga postures (*Asana*) for 15 min followed by 5 min of breathing techniques. The yoga postures included postures in standing, sitting, prone, and supine position. The breathing techniques (*Pranayama*) involved voluntary regulation of breathing like breathing with forceful exhalation and passive inhalation, breathing with rapid inhalation and exhalation, as well as slow and rhythmic alternate nostril breathing. At the end of each session yogic games for memory, awareness and creativity were completed (Figure 2). Throughout the different phases of the yoga session, a story was told to motivate the children for an active participation. The level of difficulty of yoga sessions increased from week to week to match the children's adaptations.

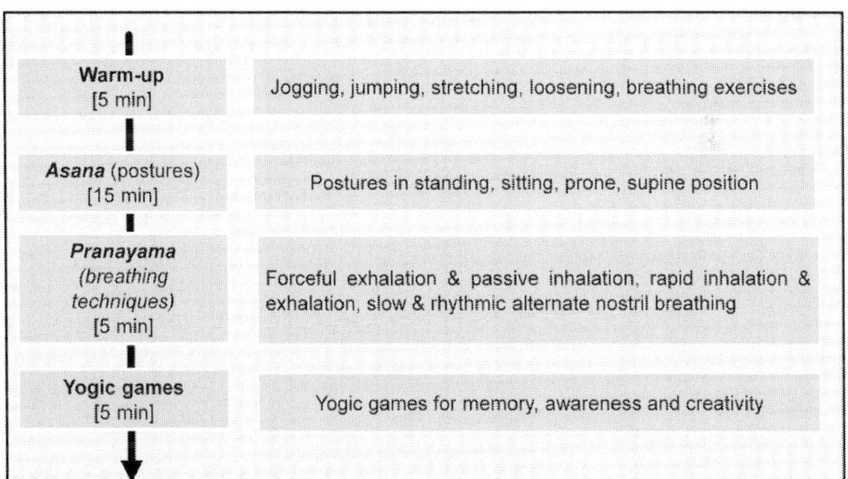

Figure 2. Details of the yoga program.

The PE group performed 24 regular PE sessions (two PE sessions per week, 30 min per session with moderate intensity) involving game-based activities, like basketball, soccer, handball, and relay games. The PE program was conducted by an instructor (trained undergraduate student) in the Kindergarten's gym.

In order to blind the interventions the following procedure was applied during the intervention period: Both yoga and PE were instructed and supervised by individuals not belonging to the Kindergarten's staff. The participating children were picked up in their respective groups by the investigators and escorted to the respective Gyms where yoga and PE were performed. The children were left in the custody of the yoga teacher, respectively, with the undergraduate student for the PE group. The children randomized to the passive CG were escorted by the investigators to other group rooms of the Kindergarten, which did not belong to the class of the children. In that group rooms, the children were free to play and engage in self-chosen activities like free play or artisanal activities. These group rooms were supervised by other Kindergarten teachers not belonging to the class of the child. All three activities (yoga, PE, and CG) were conducted parallel at the same time and in the absence of the teachers who rated the ADHD relevant behavior pretest and posttest.

Participants attended 91% of PE sessions and 88% of yoga sessions.

The CG completed no kind of physical exercise during the duration of the study.

Visual Attention Test

The visual attention test is one out of 27 subtest of the Developmental Neuropsychological Assessment (NEPSY) (Korkman et al., 1998). An instrument that provides a comprehensive neuropsychological assessment grouped into five domains (attention/executive function, language, memory and learning, sensorimotor functions, and visuospatial processing) for children aged 3–12 years (Korkman et al., 1998). Briefly, the Visual Attention subtest is allocated to the domain Attention and Executive Functioning and comprises a two-part visual cancellation task that assesses speed and precision at detecting targets among distractors.

Visuomotor Precision Test

The *NEPSY–Second Edition* (NEPSY-II) (Korkman et al., 2007) is an updated and modified version of the NEPSY (Korkman et al., 1998) and assessing neuropsychological functioning across the six domains attention and executive functioning, language, memory and learning, sensorimotor, social perception, and visuospatial processing with the help of several subtests. As part of the NEPSY-II, as described in detail elsewhere (Korkman et al., 2007), the subtest *Visuomotor Precision* is designed to assess neuropsychological functioning in the domain of sensorimotor function. The subtest measured the time in which participants perform a visuomotor task as well as the precision of the graphics performed in this task. The children using the preferred hand to draw lines inside of tracks as quickly and precise as possible.

ADHD Rating Scale-IV

The ADHD Rating Scale-IV (DuPaul et al., 1998) is a norm-referenced checklist that measures the symptoms of ADHD based on the diagnostic criteria of the *Diagnostic and Statistical Manual of Mental Disorders* (American Psychiatric Association [APA], 2000). As described in detail elsewhere (DuPaul et al., 1998), the ADHD Rating Scale-IV (French version) measures the frequency of two distinct behaviors: (1) symptoms associated with inattention and (2) symptoms associated with hyperactivity/impulsivity. In the present study, the inattention and hyperactivity behaviors of the participants were evaluated by the Kindergarten teacher with the help of the school version of the ADHD Rating Scale-IV (DuPaul et al., 1998). Whereas, the same preschool teacher evaluated the child at (T_0) and (T_1). The teacher assesses symptom frequency using a four-point Likert scale. The questionnaire consists of 18 items, nine items relating to inattention and nine items relating to hyperactivity. The behavioral scale of the students (DuPaul et al., 1998) is an instrument for assessing frequency of inattention and hyperactivity/impulsivity behaviors. Results are described in terms of the Inattention subscale and the Hyperactivity-impulsivity subscale as well as the total score.

Empirical Hypothesis

Yoga has a significant positive impact on the development of visual attention, visuomotor precision, and ADHD behavior in comparison to generic PE as well as in comparison to CG.

Statistical Analysis

Statistical tests were processed using SPSS software for Windows, version 25 (IBM, Armonk, NY, United States). Mean and standard deviation (SD) values were calculated for each variable. G power software (Faul et al.,

2007) was applied to calculate the required sample size. Values for α were set at 0.05 and power at 0.80. Based on discussions between the authors, effect size was estimated to be 0.7 (medium effect). Required sample size was 55 participants. Due to practical reasons like availability, 45 children could be included.

The Shapiro-Wilk test revealed that data of T_0 and T_1 was normally distributed and homogeneity of variance (Levene test) was given with no transformation necessary. A repeated measures analysis of variance was performed to analyze the impact of yoga in comparison to generic PE (2 [group] × 2 [time-of-measure]) as well as in comparison to CG (2 [group] × 2 [time-of-measure]). In addition to repeated measures analysis of variance, Bonferroni–Holm corrections were performed. Furthermore, data of T_0 were checked for significant differences between the three groups, applying a one-way ANOVA. A probability level of 0.05 was selected a priori as the criterion for statistical significance. To estimate practical relevance, effect sizes (partial eta squared, ηp^2) (Richardson, 2011) were calculated with $\eta p^2 \geq 0.01$ indicated small, ≥0.06 medium, and ≥0.14 large effects (Clark-Carter, 1997).

RESULTS

All parameters measured at T_0 and T_1 with accompanying statistical analyses are presented in Table 1. All parameters measured at T_0 did not show significant differences between the three groups ($p > 0.001$).

Table 1. Parameters of NEPSY, NEPSY-II and ADHD Rating Scale-IV (means ± SD) for the control group ($n = 15$), physical education group ($n = 15$), and yoga group ($n = 15$) before (T_0) and after (T_1) the 12-week intervention.

		T_0	T_1	Time effect				Group effect				Time × group effect			
				$F(1,28)$	p	Adj. p	ηp^2	$F(1,28)$	p	Adj. p	ηp^2	$F(1,28)$	P	Adj. p	ηp^2
Attention A [min]	Yoga	1.41 ± 0.07	1.31 ± 0.05												
	PE	1.41 ± 0.59	1.35 ± 0.05	28.38	0.000	0.007	0.503	1.80	0.191	0.025	0.060	2.62	0.117	0.050	0.086
	Control	1.41 ± 0.06	1.39 ± 0.07	12.58	0.001	0.017	0.310	8.55	0.007	0.005	0.234	5.23	0.030	0.008	0.157
Attention B [score]	Yoga	8.87 ± 1.64	12.20 ± 1.61												
	PE	8.93 ± 1.03	10.53 ± 1.30	44.47	0.000	0.004	0.614	4.86	0.036	0.008	0.148	5.49	0.026	0.007	0.164
	Control	9.33 ± 1.80	10.13 ± 1.68	30.47	0.000	0.006	0.521	2.68	0.113	0.010	0.087	11.45	0.002	0.005	0.290
VM precision A [s]	Yoga	24.58 ± 1.25	21.06 ± 1.47												
	PE	24.25 ± 1.17	23.69 ± 1.86	37.29	0.000	0.005	0.571	7.63	0.010	0.006	0.214	19.62	0.000	0.004	0.412
	Control	26.17 ± 2.09	24.64 ± 3.21	31.48	0.000	0.005	0.529	16.30	0.000	0.004	0.368	4.92	0.035	0.010	0.149

VM precision A [score]	Yoga	11.33 ± 1.45	8.93 ± 1.75												
	PE	11.73 ± 1.91	11.07 ± 1.22	15.08	0.001	0.010	0.350	8.57	0.007	0.005	0.234	4.82	0.037	0.013	0.147
	Control	11.93 ± 2.08	11.13 ± 2.45	10.97	0.003	0.025	0.282	6.92	0.014	0.006	0.198	2.74	0.109	0.025	0.089
VM precision B [min]	Yoga	1.51 ± 0.07	1.33 ± 0.08												
	PE	1.49 ± 0.07	1.46 ± 0.07	82.26	0.000	0.003	0.746	5.80	0.023	0.007	0.172	34.96	0.000	0.003	0.555
	Control	1.47 ± 0.12	1.44 ± 0.14	14.63	0.001	0.013	0.343	1.44	0.240	0.050	0.049	7.64	0.010	0.006	0.214
VM precision B [score]	Yoga	21.93 ± 2.37	18.67 ± 1.35												
	PE	21.87 ± 2.45	20.47 ± 1.56	29.22	0.000	0.006	0.511	2.23	0.147	0.013	0.074	4.68	0.039	0.017	0.143
	Control	21.93 ± 2.09	22.07 ± 2.19	6.35	0.018	0.050	0.185	17.39	0.000	0.004	0.383	7.48	0.011	0.006	0.211
Hyperactivity	Yoga	13.93 ± 1.67	9.60 ± 1.59												
	PE	13.67 ± 1.59	11.27 ± 1.28	134.22	0.000	0.003	0.827	2.11	0.157	0.017	0.070	11.07	0.002	0.005	0.283
	Control	14.47 ± 1.96	13.20 ± 1.93	58.66	0.000	0.004	0.677	14.41	0.001	0.004	0.340	17.59	0.000	0.004	0.386

Inattention	Yoga	12.40 ± 1.50	6.87 ± 1.64												
	PE	12.73 ± 1.53	10.27 ± 0.96	100.70	0.000	0.003	0.782	30.20	0.000	0.003	0.519	14.80	0.001	0.004	0.346
	Control	12.60 ± 1.99	12.73 ± 3.15	24.35	0.000	0.004	0.465	27.98	0.000	0.003	0.500	26.82	0.000	0.003	0.489
Total	Yoga	26.33 ± 2.19	16.47 ± 2.33												
	PE	26.40 ± 2.50	21.53 ± 1.55	211.56	0.000	0.003	0.883	17.66	0.000	0.003	0.387	24.37	0.000	0.003	0.465
	Control	27.07 ± 3.06	25.93 ± 4.57	52.13	0.000	0.008	0.651	33.80	0.000	0.003	0.547	32.86	0.000	0.003	0.540

N_{Yoga} = 15; N_{PE} = 15; $N_{Control}$ = 15; min = minute; s = seconds; VM precision A (s) = visuomotor precision A; VM precision B (min) = visuomotor precision B (min); total score = total score of the ADHD Rating Scale-IV (DuPaul et al., 1998); adj. P = p-values Bonferroni–Holm corrected; PE = group physical education; CG = control group with no physical activity; F = degrees of freedom; p = probability; ηp^2 = effect size partial eta-square.

Visual Attention

Yoga has a non-significant impact on the development of visual attention A in comparison to generic PE [time × group effect: $F(1,28) = 2.62$; $p = 0.117$; $\eta p^2 = 0.086$]. However, yoga has a significant positive impact on the development of visual attention A in comparison to CG [time × group effect: $F(1,28) = 5.23$; $p = 0.030$; $\eta p^2 = 0.157$]; yoga (T_0: 1.41 ± 0.07, T_1: 1.31 ± 0.05) leads to comparatively faster attention times (CG: T_0: 1.41 ± 0.06, T_1: 1.39 ± 0.07). In addition, yoga has a significant positive impact on the development of visual attention B in comparison to PE [time × group effect: $F(1,28) = 5.49$; $p = 0.026$; $\eta p^2 = 0.164$] as well as in comparison to CG [time × group effect: $F(1,28) = 11.45$; $p = 0.002$; $\eta p^2 = 0.290$]; in both cases, yoga (T_0: 8.87 ± 1.64, T_1: 12.20 ± 1.61) leads to comparatively higher scores (PE: T_0: 8.93 ± 1.03, T_1: 10.53 ± 1.30; CG: T_0: 9.33 ± 1.80, T_1: 10.13 ± 1.68). All significant interaction effects can be classified as large.

However, after Bonferroni–Holm correction, only one significant time × group interaction effect remained. Hereafter, yoga has a significant positive impact on the development of visual attention B in comparison to CG (adj. $p = 0.005$).

Visuomotor Precision

Yoga has a significant positive impact on the development of visuomotor precision A in comparison to generic PE [time × group effect: $F(1,28) = 19.62$; $p = 0.000$; $\eta p^2 = 0.412$] as well as in comparison to CG [time × group effect: $F(1,28) = 4.92$; $p = 0.035$; $\eta p^2 = 0.149$]; in both cases, yoga (T_0: 24.58 ± 1.25, T_1: 21.06 ± 1.47) leads to comparatively faster visuomotor precision times (PE: 24.25 ± 1.17, T_1: 23.69 ± 1.86; CG: T_0: 26.17 ± 2.09, T_1: 24.64 ± 3.21). Further, yoga has a significant positive impact on the development of visuomotor precision A in comparison to generic PE [time × group effect: $F(1,28) = 4.82$; $p = 0.037$; $\eta p^2 = 0.147$]; yoga (T_0: 11.33 ± 1.45, T_1: 8.93 ± 1.75) leads to comparatively lower error rates (PE: T_0: 11.73 ± 1.91, T_1: 11.07 ± 1.22). However, yoga has no significant impact on the development of visual attention A in comparison to CG [time × group effect: $F(1,28) = 2.74$; $p = 0.109$; $\eta p^2 = 0.089$]. In addition, yoga has a significant positive impact on the development of visuomotor precision B in comparison to generic PE [time × group effect: $F(1,28) = 34.96$; $p = 0.000$; $\eta p^2 = 0.555$] as well as in comparison to CG [time × group effect: $F(1,28) = 7.64$; $p = 0.010$; $\eta p^2 = 0.214$]; in both cases, yoga (T_0: 1.51 ± 0.07, T_1: 1.33 ± 0.08) leads to comparatively faster visuomotor precision times (PE: T_0: 1.49 ±

0.07, T_1: 1.46 ± 0.07; CG: T_0: 1.47 ± 0.12, T_1: 1.44 ± 0.14). Finally, yoga has a significant positive impact on the development of visuomotor precision B in comparison to generic PE [time × group effect: $F(1,28) = 4.68$; $p = 0.039$; $\eta p^2 = 0.143$] as well as in comparison to non-treatment conditions [time × group effect: $F(1,28) = 7.48$; $p = 0.011$; $\eta p^2 = 0.211$]; in both cases, yoga (T_0: 21.93 ± 2.37, T_1: 18.67 ± 1.35) leads to comparatively lower error rates (PE: T_0: 21.87 ± 2.45, T_1: 20.47 ± 1.56; CG: T_0: 21.93 ± 2.09, T_1: 22.07 ± 2.19). All significant interaction effects can be classified as large.

However, after Bonferroni–Holm correction, only two significant time × group interaction effects remained. Hereafter, yoga has a significant positive impact on both, the development of visuomotor precision A (completion time) in comparison to PE (adj. $p = 0.004$) as well as visuomotor precision B (completion time) in comparison to PE group (adj. $p = 0.003$).

ADHD Rating Scale-IV

Yoga has a significant positive impact on the development of hyperactivity behavior in comparison to generic PE [time × group effect: $F(1,28) = 11.07$; $p = 0.002$; $\eta p^2 = 0.283$] as well as in comparison to CG [time × group effect: $F(1,28) = 17.59$; $p = 0.000$; $\eta p^2 = 0.386$]; in both cases, yoga (T_0: 13.93 ± 1.67, T_1: 9.60 ± 1.59) leads to comparatively lower hyperactivity behavior (PE: 13.67 ± 1.59, T_1: 11.27 ± 1.28; CG: T_0: 14.47 ± 1.96, T_1: 13.20 ± 1.93). Further, yoga has a significant positive impact on the development of inattention behavior in comparison to generic PE [time × group effect: $F(1,28) = 14.80$; $p = 0.001$; $\eta p^2 = 0.346$] as well as in comparison to CG [time × group effect: $F(1,28) = 26.82$; $p = 0.000$; $\eta p^2 = 0.489$]; in both cases, yoga (T_0: 12.40 ± 1.50, T_1: 6.87 ± 1.64) leads to comparatively lower inattention scores (PE: 12.73 ± 1.53, T_1: 10.27 ± 0.96; CG: T_0: 12.60 ± 1.99, T_1: 12.73 ± 3.15). Finally, yoga has a significant positive impact on the ADHD total score in comparison to generic PE [time × group effect: $F(1,28) = 24.37$; $p = 0.000$; $\eta p^2 = 0.465$] as well as in comparison to CG [time × group effect: $F(1,28) = 32.86$; $p = 0.000$; $\eta p^2 = 0.540$]; in both cases, yoga (T_0: 26.33 ± 2.19, T_1: 16.47 ± 2.33) leads to comparatively lower ADHD-total-scores (PE: 26.40 ± 2.50, T_1: 21.53 ± 1.55; CG: T_0: 27.07 ± 3.06, T_1: 25.93 ± 4.57). All significant interaction effects can be classified as large. Following Bonferroni–Holm corrections, all reported time × group interaction effects remained significant (Table 1).

DISCUSSION

The aim of this study was to investigate the effects of 12-week Kindergarten-based yoga practice on visual attention, visual-motor precision, as well as on hyperactivity and inattention behavior in 5-year-old Tunisian children.

The main findings of the present study were as follows:

(1) Yoga had a significant positive effect over the 12-week intervention period on the majority of parameters of visual attention and on parameters of visual-motor precision, in comparison to the PE and CG. However, after Bonferroni–Holm correction, yoga only had a significant positive impact on the development of visual attention B in comparison to CG. Considering visual-motor precision after Bonferroni–Holm correction, yoga only had a significant positive impact on the development of visuomotor precision A (completion time) in comparison to PE, as well as on visuomotor precision B (completion time) in comparison to PE group.

(2) Hyperactivity and inattention behavior and the total score of the ADHD Rating Scale-IV improved significantly in the yoga group from T_0 to T_1, in comparison to the PE and CG.

The results demonstrate that children participating in the yoga program, in comparison to the PE and CG, improved their skills related to cognitive functions, measured by the subtests of NEPSY (Korkman et al., 1998) and NEPSY-II (Korkman et al., 2007), as well as their behavior in respect to inattention and hyperactivity, evaluated by the ADHD Rating Scale-IV (DuPaul et al., 1998). The amelioration of attention with the help of yoga in the present study is in line with previous results demonstrating that sensorimotor training enhances attention in children with (Banaschewski et al., 2001; Cohen et al., 2018) and without ADHD (Mak et al., 2018). Our results are also coherent with previous studies reporting that school-based prevention programs focusing on yoga or sensorimotor training for young children can reduce deficits in attention and improve academic performance (Paour and Cèbes, 2001; Chevalier and Simard, 2006; Krisanaprakornkit et al., 2010; Gothe et al., 2013). In addition, our results support findings of previous research demonstrating that yoga practice decreases ADHD symptoms in preschool children (Cohen et al., 2018). Although, no manifest ADHD diagnosis (e.g., from a clinician, psychologist, or pediatrician) was evident for participants of the present study, many participants showed symptoms of inattention, hyperactivity, and impulsivity at T_0 (Table 1).

With the yoga intervention, the inattention behavior in the present study was reduced effectively.

Since attention represents an important prerequisite in the processes of learning (Flessas and Lussier, 2001; McClelland et al., 2013) and is associated with better academic performance in preschool children (Duncan et al., 2007) as well as executive functions (Friedman et al., 2007) and academic performance (McClelland et al., 2013) in the adolescence, the present study demonstrated that yoga has the potential to facilitate learning processes and may contribute to academic achievement of preadolescent children. Furthermore, it seems to be important to work on attention problems at an early age, because attention problems during childhood seem to be associated with poorer executive functions in late adolescence (Friedman et al., 2007). Recent reviews calculated a positive small to moderate effect on attention for acute physical activity and a large positive effect on attention for chronic physical activity (de Greeff et al., 2018). In agreement with the recent review (de Greeff et al., 2018), the present study revealed large positive effect sizes for the chronic yoga intervention on attention. It is suggested that the positive effects on attention leading to increased time children are engaged in academic tasks, which has the potential of better academic performance in the long run (Duncan et al., 2007; Oberer et al., 2018). It is hypothesized that both increased physical activity, respectively, an increased physical fitness as a result of the latter leading to an increased activity in selective structures of the brain, causing functional and structural connectivity (Esteban-Cornejo et al., 2017). As a result, physical activity potentially facilitates factors of cognitive performance and executive functions (Diamond, 2000; Colcombe and Kramer, 2003). For the exact underlying mechanisms of structural and functional brain changes associated with physical exercise in children, the available research is very limited in number and scope. Whereas, associations of cardiorespiratory fitness and speed-agility in obese children (8–11 years) with greater gray matter volumes, which in turn reflect children's academic performance, could be demonstrated recently (Esteban-Cornejo et al., 2017). Additionally, an 8-month aerobic exercise program with 8–11-year-old obese children caused improved white matter integrity in the brain tract connecting frontal and temporal lobes (Schaeffer et al., 2014). In addition, an improved memory performance in children was associated with improved fronto-temporal circuitry as a result of aerobic exercise (Erickson et al., 2011). To the best of our knowledge, no studies are available on yoga and the effects of structural alterations of brain matter in children. Whereas, reviewed data suggest several structural alterations and activation changes of brain structures (e.g., amygdala, frontal lobes, white matter) in adults as a result of yoga (for detailed description of specific changes, see Desai et al., 2015). Similar alterations in structural and functional capacities in the brain of children may be responsible for the alterations of attention and behavior in the present study.

The present results imply that improvements in cognitive tasks were achieved after a yoga intervention compared to generic PE and the usual Kindergarten activities as represented in the CG. These results are in agreement with previous studies, which reported that young yoga participants showed significant differences over time on measures of executive functions (Manjunath and Telles, 2004; Chaya et al., 2012; Purohit and Pradhan, 2016). Likewise, Telles et al. (1993) revealed that children, practicing yoga for 10 days, improved in spatial memory scores, strategic planning, and concentration. In the same context, a significant improvement following 12 weeks of school-based yoga was observed for cognitive functions and memory in 13-year-old children (Verma et al., 2014). In agreement with studies on yoga with older children, our data demonstrate that preschool-based yoga is a suitable physical exercise to support cognitive abilities of preschool children. Since the level of physical activity, executive functions, and visual-motor coordination in Kindergarten children seem to be associated with later academic achievement (Oberer et al., 2018), it seems to be beneficial to implement regular yoga classes at preschools to contribute to the basis for academic achievement in young children. In this context, several authors advocate the potential of yoga for schools as an enjoyable and cost-effective exercise to improve cognitive and behavioral skills, which are relevant for school and may contribute to academic achievement (Kauts and Sharma, 2009; Chaya et al., 2012; Verma et al., 2014). The observation of the practical application of yoga at schools worldwide reveals that yoga is increasingly important in the school setting (Flak, 2003; Chaya et al., 2012) since it is recognized in the curriculum in France, Brazil, Canada, and Italy (Flak, 2003). Additionally, in Italy, classroom-based yoga is performed in all schools since 2000 (Flak, 2003).

Although no explicit ADHD patients were among participants of the present study, the yoga intervention reduced the behavior of hyperactivity and inattention significantly and to a practical relevant extent, which is in line with numerous studies. Previous research demonstrated that a sensorimotor training program for children leads to the reduction of behavioral problems (Eggert and Lütje, 1991) and ameliorates anger control and anxiety (Khalsa et al., 2012), factors that potentially interfere with academic achievement and psychosocial well-being. Furthermore, classroom-based yoga leads to a significant improvement in emotional and psychosocial quality of life and reduced anxiety in third-grade students (Bazzano et al., 2018). Additionally, Noggle et al. (2012) reported that school-based yoga in high school students caused increases in measures of psychosocial well-being in comparison with the regular PE. Consequently, yoga could be considered as a method to teach coping skills for young students which potentially contribute to mental health and psychosocial well-being.

As noted by Rossner (1995), yoga would allow relaxation to improve attention to task and spatial relationships in the environment, which we confirmed by our results. The findings of the present study provide evidence for the importance of practicing yoga at Kindergarten as it helps reach higher cognitive performance and consequently potentially higher academics performances in children. Thus, yoga seems to be an appropriate tool to a cognitive education to preschool Tunisian children as a complement to the usual social, emotional, and artisanal learning activities. As recommended (Paour and Cèbes, 2001), yoga would allow the preschool children to effectively engage attentional abilities and improve the executive self-regulation of the action.

Limitations

Whereas, it would be useful to have a larger sample since our study consisted of 45 participants and this may present a limitation since the variability may be significant in this sample. Furthermore, a greater number of participants would have given more statistical power to the data interpretation with less risk of calculating type 2 errors. At the same time, the small sample size allowed us to increase availability, motivation, and compliance of children during participation.

The duration of a yoga session is usually approximately 1:20 h (Lark, 2003) and the 30 min of yoga allotted to each class in the present study could represent a limitation. We reduced the sessions to 30 min in order to respect the attention span of the children as well as the schedule allocated for the daily routines of the Kindergarten. Nevertheless, the duration of 30 min of yoga sessions in the present study is within the usual range of 30–40 min of comparable studies (Chou and Huang, 2017; Bazzano et al., 2018).

The inattention and hyperactivity behaviors of the participants were evaluated by the Kindergarten teacher with the help of the ADHD Rating Scale-IV (DuPaul et al., 1998). The Kindergarten teachers were not informed about the hypothesis of the study and blinded to the intervention groups and the passive CG. We only informed Kindergarten teachers that different kinds of physical exercise were tested on cognitive functions in the children. Whereas, we cannot exclude that children told their Kindergarten teachers about the performed intervention or that Kindergarten teachers eventually had intuitions concerning expected results of the yoga intervention.

The present study is part of a research project focusing on graphic design and its link with motor and cognitive aspects. We consider fine motor skills as especially important during preschool age since writing is a fine motor activity that is initiated during the preschool period and requires perceptual-motor skills. In children aged between 2–4 years, the vision serves as a primary guide for graphic activities like writing (James, 2010; Maldarelli et al., 2015). Therefore, we did not include auditory attention tests. Since auditory stimuli are an additional prerequisite for graphic activities, we are looking forward to implement auditory attention tests in the following parts of the research project.

CONCLUSION

Twelve weeks of two 30-min yoga sessions per week improved certain parameters of attention, visual-motor precision, and reduced behavior of inattention and hyperactivity in 5-year-old Kindergarten children. The yoga program improves attention, visual-motor precision, and behavior to a higher extent compared to generic PE and the usual Kindergarten activities as represented in the passive CG. As a consequence, Kindergarten-based yoga classes represents a sufficient and cost-benefit effective exercise activity which may enhance functions relevant for learning among young children and could be added as a complement to social, emotional, and artisanal learning activities at the Kindergarten.

ETHICS STATEMENT

In this single-center, three-arm randomized, controlled study, 45 healthy children (28 female; 17 male; 5.2 ± 0.4 years) of a private Tunisian Kindergarten volunteered to participate. All participants and their legal guardians were informed in detail, in written form as well as orally, about the design of the study, including the potential risks and benefits, before providing their informed written consent to participate. Participants were free to withdraw from the study at any time without further consequences.

The inclusion criteria were a lack of any frequent participation in yoga exercise programs for at least 6 months prior to the study; no daily intake of medication; and for inclusion in the analysis, completion of at least 80% of the yoga sessions.

All procedures were conducted in accordance with the Code of Ethics for human experimentation of the World Medical Association, the Declaration of Helsinki (World Medical Association, 2013), as well as the ethical standards of the International Journal of Sports Medecine (Harriss and Atkinson, 2015). The experimental protocol was pre-approved by the ethical review board of the High Institute of Sport and Physical Education of University of Sfax, Tunisia.

AUTHOR CONTRIBUTIONS

All authors have made a substantial, direct, and intellectual contribution to the work and approved it for publication.

FUNDING

This study was financially supported by the Ministry of Higher Education and Scientific Research, Tunisia. We acknowledge financial support by Deutsche Forschungsgemeinschaft within the funding programme Open Access Publishing, by the Baden-Württemberg Ministry of Science, Research, and the Arts, and by Ruprecht-Karls-Universität Heidelberg.

CONFLICT OF INTEREST STATEMENT

The authors declare that the research was conducted in the absence of any commercial or financial relationships that could be construed as a potential conflict of interest.

ACKNOWLEDGMENTS

The authors wish to express their sincere gratitude to all the participants for their maximal effort and cooperation.

REFERENCES

1. Alesi, M., Bianco, A., Padulo, J., Vella, F. P., Petrucci, M., Paoli, A., et al. (2014). Motor and cognitive development: the role of karate. *Muscle Ligaments Tendons J.* 14, 114–120. doi: 10.11138/mltj/2014.4.2.114
2. Altenburg, T. M., Chinapaw, M. J., and Singh, A. S. (2016). Effects of one versus two bouts of moderate intensity physical activity on selective attention during a school morning in Dutch primary schoolchildren: a randomized controlled trial. *J. Sci. Med. Sport.* 19, 820–824. doi: 10.1016/j.jsams.2015.12.003
3. American Psychiatric Association [APA] (2000). *DSM-IV-TR: Diagnostic and Statistical Manual of Mental Disorders.* Washington, DC: American Psychiatric Association.
4. American Psychiatric Association [APA] (2013). *Diagnostic and Statistical Manual of Mental Disorders*, 5th Edn. Washington, DC: American Psychiatric Association. doi: 10.1176/appi.books.9780890425596
5. Balasubramaniam, M., Telles, S., and Doraiswamy, P. M. (2012). Yoga on our minds: a systematic review of yoga for neuropsychiatric disorders. *Front. Psychiatry* 25:117. doi: 10.3389/fpsyt.2012.00117
6. Banaschewski, T., Besmens, F., Zieger, H., and Rothenberger, A. (2001). Evaluation of sensorimotor training in children with ADHD. *Percept. Mot. Skills* 92, 137–149. doi: 10.2466/pms.2001.92.1.137
7. Bazzano, A. N., Anderson, C. E., Hylton, C., and Gustat, J. (2018). Effect of mindfulness and yoga on quality of life for elementary school students and teachers: results of a randomized controlled school-based study. *Psychol. Res. Behav. Manag.* 11, 81–89. doi: 10.2147/PRBM.S157503
8. Blair, C., and Raver, C. C. (2015). School readiness and self-regulation: a developmental psychobiological approach. *Annu. Rev. Psychol.* 66, 711–731. doi: 10.1146/annurev-psych-010814-015221
9. Brethenoux-Seguin, F. (2007). *Yoga & Bonne Santé.* Paris: Edition ESTEM De Boeck.
10. Chaya, M. S., Nagendra, H., Selvam, S., Kurpad, A., and Srinivasan, K. (2012). Effect of yoga on cognitive abilities in schoolchildren from a socioeconomically disadvantaged background: a randomized controlled study. *J. Altern. Complement. Med.* 18, 1161–1167. doi: 10.1089/acm.2011.0579

11. Chevalier, N., and Simard, F. (2006). "Programmes sensorimoteurs pour jeunes ayant un TDAH et des troubles associés," in *Trouble Déficitaire De L'attention Avec Hyperactivité: Soigner, Éduquer, Surtout Valoriser*, ed. S. Foy (Québec: Presses de l'Université du Québec), 141–158.
12. Chou, C. C., and Huang, C. J. (2017). Effects of an 8-week yoga program on sustained attention and discrimination function in children with attention deficit hyperactivity disorder. *PeerJ.* 5:e2883. doi: 10.7717/peerj.2883
13. Clark-Carter, D. (1997). *Doing Quantitative Psychological Research: From Design to Report.* Hove: Psychology Press.
14. Cohen, S. C. L., Harvey, D. J., Shields, R. H., Shields, G. S., Rashedi, R. N., Tancredi, D. J., et al. (2018). Effects of yoga on attention, impulsivity, and hyperactivity in preschool-aged children with attention-deficit hyperactivity disorder symptoms. *J. Dev. Behav. Pediatr.* 39, 200–209. doi: 10.1097/DBP.0000000000000552
15. Colcombe, S. J., and Kramer, A. F. (2003). Fitness effects on the cognitive function of older adults: a meta-analytic study. *Psychol. Sci.* 14, 125–130. doi: 10.1111/1467-9280.t01-1-01430
16. Danielson, M. L., Bitsko, R. H., Ghandour, R. M., Holbrook, J. R., Kogan, M. D., and Blumberg, S. J. (2018). Prevalence of parent-reported ADHD diagnosis and associated treatment among U.S. children and adolescents, 2016. *J. Clin. Child. Adolesc. Psychol.* 47, 199–212. doi: 10.1080/15374416.2017.1417860
17. de Greeff, J. W., Bosker, R. J., Oosterlaan, J., Visscher, C., and Hartman, E. (2018). Effects of physical activity on executive functions, attention and academic performance in preadolescent children: a meta-analysis. *J. Sci. Med. Sport* 21, 501–507. doi: 10.1016/j.jsams.2017.09.595
18. Delvolvé, N. (2005). *Tous Les Élèves Peuvent Apprendre. Approches Psychologiques Et Ergonomiques Des Apprentissages Scolaires.* Paris: Ed. Hachette Education.
19. Desai, R., Tailor, A., and Bhatt, T. (2015). Effects of yoga on brain waves and structural activation: a review. *Complement. Ther. Clin. Pract.* 21, 112–118. doi: 10.1016/j.ctcp.2015.02.002
20. Diamond, A. (2000). Close interrelation of motor development and cognitive development and of the cerebellum and prefrontal cortex. *Child Dev.* 71, 44–56. doi: 10.1111/1467-8624.00117

21. Diamond, A. (2012). Activities and programs that improve children's executive functions. *Curr. Dir. Psychol. Sci.* 21, 335–341. doi: 10.1177/0963721412453722
22. Diamond, A. (2013). Executive functions. *Annu. Rev. Psychol.* 64, 135–168. doi: 10.1146/annurev-psych-113011-143750
23. Duncan, G. J., Dowsett, C. J., Claessens, A., Magnuson, K., Huston, A. C., Klebanov, P., et al. (2007). School readiness and later achievement. *Dev. Psychol.* 43, 1428–1446. doi: 10.1037/0012-1649.43.6.1428
24. DuPaul, G. J., Power, T. J., Anastopoulos, A. D., and Reid, R. (1998). *ADHD Rating Scale-IV: Checklists, Norms and Clinical Interpretation.* New York, NY: Guilford.
25. Eggert, D., and Lütje, B. (1991). Psychomotorik in der (Sonder-)Schule. Empirische studien zu den grenzen eines förderkonzepts. *Praxis Der Psychomotorik* 3, 156–168.
26. Erickson, K. I., Voss, M. W., Prakash, R. S., Basak, C., Szabo, A., Chaddock, L., et al. (2011). Exercise training increases size of hippocampus and improves memory. *Proc. Natl. Acad. Sci. U.S.A.* 108, 3017–3022. doi: 10.1073/pnas.1015950108
27. Esteban-Cornejo, I., Cadenas-Sanchez, C., Contreras-Rodriguez, O., Verdejo-Roman, J., Mora-Gonzalez, J., Migueles, J. H., et al. (2017). A whole brain volumetric approach in overweight/obese children: examining the association with different physical fitness components and academic performance. The active brains project. *Neuroimage* 159, 346–354. doi: 10.1016/j.neuroimage.2017.08.011
28. Faul, F., Erdfelder, E., Lang, A. G., and Buchner, A. G. (2007). G□Power 3: a flexible statistical power analysis program for the social, behavioral, and biomedical sciences. *Behav. Res. Methods* 2007, 175–191. doi: 10.3758/BF03193146
29. Fedewa, A. L., and Ahn, S. (2011). The effects of physical activity and physical fitness on children's achievement and cognitive outcomes: a meta-analysis. *Res. Q. Exerc. Sport* 82, 521–535. doi: 10.1080/02701367.2011.10599785
30. Flak, M. (2003). *Recherche Sur Le Yoga Dans L'éducation. 3ème Millénaire: Spiritualité – Connaissance De Soi – Non-Dualité – Méditation, 125*. Available at: http://www.rye-yoga.fr/ (accessed July 15, 2018).

31. Flessas, L., and Lussier, F. (2001). *Neuropsychologie De L'enfant. Troubles Développementaux Et De L'apprentissage.* Paris: Dunod.
32. Folleto, J. C., Pereira, K. R., and Valentini, N. C. (2016). The effects of yoga practice in school physical education on children's motor abilities and social behavior. *Int. J. Yoga* 9, 156–162. doi: 10.4103/0973-6131.183717
33. Friedman, N. P., Haberstick, B. C., Willcutt, E. G., Miyake, A., Young, S. E., Corley, R. P., et al. (2007). Greater attention problems during childhood predict poorer executive functioning in late adolescence. *Psychol. Sci.* 18, 893–900. doi: 10.1111/j.1467-9280.2007.01997.x
34. Galantino, M. L., Galbavy, R., and Quinn, L. (2008). Therapeutic effects of yoga for children: a systematic review of the literature. *Pediatr. Phys. Ther.* 20, 66–80. doi: 10.1097/PEP.0b013e31815f1208
35. Gothe, N., Pontifex, M. B., Hillman, C., and McAuley, E. (2013). The acute effects of yoga on executive function. *J. Phys. Act. Health* 10, 488–495. doi: 10.1123/jpah.10.4.488
36. Haapala, E. (2012). Physical activity, academic performance and cognition in children and adolescents. A systematic review. *Baltic J. Health Phys. Act.* 4, 53–61. doi: 10.2478/v10131-012-0007-y
37. Harrison, L. J., Manocha, R., and Rubia, K. (2004). Sahaja yoga meditation as a family treatment programme for children with attention deficit-hyperactivity disorder. *Clin. Child Psychol. Psychiatry* 9, 479–497. doi: 10.1177/1359104504046155
38. Harriss, D. J., and Atkinson, G. (2015). Ethical standards in sport and exercise science research: 2016 update. *Int. J. Sports Med.* 36, 1121–1124. doi: 10.1055/s-0035-1565186
39. Hillman, C. H., Pontifex, M. B., Castelli, D. M., Khan, N. A., Raine, L. B., Scudder, M. R., et al. (2014). Effects of the FITKids randomized controlled trial on executive control and brain function. *Pediatrics* 134, e1063–e1071. doi: 10.1542/peds.2013-3219
40. Innes, K. E., Bourguignon, C., and Taylor, A. G. (2005). Risk indices associated with the insulin resistance syndrome, cardiovascular disease, and possible protection with yoga: a systematic review. *J. Am. Board Fam. Pract.* 18, 491–519. doi: 10.3122/jabfm.18.6.491
41. James, K. H. (2010). Sensori-motor experience leads to changes in visual processing in the developing brain. *Dev. Sci.* 13, 279–288. doi: 10.1111/j.1467-7687.2009.00883.x

42. Jensen, P. S., and Kenny, D. T. (2004). The effects of yoga on the attention and behavior of boys with Attention-Deficit/hyperactivity Disorder (ADHD). *J. Atten. Disord.* 7, 205–216. doi: 10.1177/108705470400700403
43. Kaunhoven, R. J., and Dorjee, D. (2017). How does mindfulness modulate self-regulation in pre-adolescent children? An integrative neurocognitive review. *Neurosci. Biobehav. Rev.* 74, 163–184. doi: 10.1016/j.neubiorev.2017.01.007
44. Kauts, A., and Sharma, N. (2009). Effect of yoga on academic performance in relation to stress. *Int. J. Yoga* 2, 39–43. doi: 10.4103/0973-6131.53860
45. Khalsa, S. B., Hickey-Schultz, L., Cohen, D., Steiner, N., and Cope, S. (2012). Evaluation of the mental health benefits of yoga in a secondary school: a preliminary randomized controlled trial. *J. Behav. Health Serv. Res.* 39, 80–90. doi: 10.1007/s11414-011-9249-8
46. Korkman, M., Kirk, U., and Kemp, S. (1998). *NEPSY: A Developmental Neuropsychological Assessment Manual.* San Antonio, TX: The Psychological Corporation.
47. Korkman, M., Kirk, U., and Kemp, S. (2007). *NEPSY—Second Edition (NEPSY-II).* San Antonio, TX: Harcourt Assessment.
48. Krisanaprakornkit, T., Ngamjarus, C., Witoonchart, C., and Piyavhatkul, N. (2010). Meditation therapies for attention-deficit/hyperactivity disorder (ADHD). *Cochrane Database Syst. Rev.* 6:CD006507. doi: 10.1002/14651858.CD006507.pub2
49. Lark, L. (2003). *Yoga Pour Les Enfants.* Paris: Le Counier du Livre.
50. Leconte, C. (2005). Une si fragile attention. *Cerveau et Psychol.* 11, 36–40.
51. Lees, C., and Hopkins, J. (2013). Effect of aerobic exercise on cognition, academic achievement, and psychosocial function in children: a systematic review of randomized control trials. *Prev. Chronic Dis.* 10:E174. doi: 10.5888/pcd10.130010
52. Ludyga, S., Gerber, M., Brand, S., Holsboer-Trachsler, E., and Pühse, U. (2016). Acute effects of moderate aerobic exercise on specific aspects of executive function in different age and fitness groups: a meta-analysis. *Psychophysiology* 53, 1611–1626. doi: 10.1111/psyp.12736
53. MacCobb, S., Fitzgerald, B., Lanigan-O'Keefe, C., Irwin, N., and Mellerick, N. (2014). Students with social, emotional, and behavioural

difficulties: the Alert Program trial in post-primary schools. *J. Occup. Ther. Sch. Early Interv.* 7, 106–119. doi: 10.1080/19411243.2014.930606

54. Mak, C., Whittingham, K., Cunnington, R., and Boyd, R. N. (2018). Effect of mindfulness yoga programme MiYoga on attention, behaviour, and physical outcomes in cerebral palsy: a randomized controlled trial. *Dev. Med. Child Neurol.* 60, 922–932. doi: 10.1111/dmcn.13923
55. Maldarelli, J. E., Kahrs, B. A., Hunt, S. C., and Lockman, J. J. (2015). Development of early handwriting: visual-motor control during letter copying. *Dev Psychol.* 51, 879–888. doi: 10.1037/a0039424
56. Manjunath, N. K., and Telles, S. (2004). Spatial and verbal memory test scores following yoga and fine arts camps for school children. *Indian J. Physiol. Pharmacol.* 48, 353–356.
57. Matlin, M. (2009). *Cognition.* Hoboken, NJ: John Wiley & Sons, Inc.
58. McClelland, M. M., Acock, A. C., Piccinin, A., Rhea, S. A., and Stallings, M. C. (2013). Relations between preschool attention span-persistence and age 25 educational outcomes. *Early Child Res. Q.* 28, 314–324. doi: 10.1016/j.ecresq.2012.07.008
59. Miyake, A., Friedman, N. P., Emerson, M. J., Witzki, A. H., Howerter, A., and Wager, T. D. (2000). The unity and diversity of executive functions and their contributions to complex "frontal lobe" tasks: a latent variable analysis. *Cogn. Psychol.* 41, 49–100. doi: 10.1006/cogp.1999.0734
60. Noggle, J., Steiner, N. J., Minami, T., and Khalsa, S. B. S. (2012). Benefits of yoga for psychosocial well-being in a US high school curriculum: a preliminary randomized controlled trial jessica. *J. Dev. Behav. Pediatr.* 33, 193–201. doi: 10.1097/DBP.0b013e31824afdc4
61. Oberer, N., Gashaj, V., and Roebers, C. M. (2018). Executive functions, visual-motor coordination, physical fitness and academic achievement: longitudinal relations in typically developing children. *Hum. Mov. Sci.* 58, 69–79. doi: 10.1016/j.humov.2018.01.003
62. Ortega, F. B., Ruiz, J. R., Castillo, M. J., and Sjöström, M. (2008). Physical fitness in childhood and adolescence: a powerful marker of health. *Int. J. Obes.* 32, 1–11. doi: 10.1038/sj.ijo.0803774
63. Owen, K. B., Parker, P. D., Van Zanden, B., Macmillan, F., Astell-Burt, T., and Lonsdale, C. (2016). Physical activity and school engagement in youth: a systematic review and meta-analysis. *Educ. Psychol.* 51, 129–145. doi: 10.1080/00461520.2016.1151793

64. Paour, J. L., and Cèbes, S. (2001). L'éducation cognitive à l'école maternelle: pourquoi, comment, avec quels effets? *Revue De Psychologie De L'éducation.* 3, 86–111.

65. Peck, H. L., Kehle, T. J., Bray, M. A., and Theodore, L. A. (2005). Yoga as an intervention for children with attention problems. *Sch. Psychol. Rev.* 34, 415–424.

66. Perry, J. C., Liu, X., and Pabian, Y. (2010). School engagement as a mediator of academic performance among urban youth: the role of career preparation, parental career support, and teacher support. *Couns. Psychol.* 38, 269–295. doi: 10.1177/0011000009349272

67. Purohit, S. P., and Pradhan, B. (2016). Effect of yoga program on executive functions of adolescents dwelling in an orphan home: a randomized controlled study. *J. Tradit. Complement Med.* 7, 99–105. doi: 10.1016/j.jtcme.2016.03.001

68. Richardson, J. T. E. (2011). Eta squared and partial eta squared as measures of effect size in educational research. *Educ. Res. Rev.* 6, 135–147. doi: 10.1016/j.edurev.2010.12.001

69. Rosen, L., French, A., and Sullivan, G. (2015). Complementary, holistic, and integrative medicine. *Yoga Pediatr. Rev.* 36, 468–474. doi: 10.1542/pir.36-10-468

70. Rossner, M. Z. (1995). *Yoga, Un Nouvel Espoir Pour Les Enfants En Détresse.* Saint Zénon, QC: Louise Courteau.

71. Sandford, R., Duncombe, R., Mason, C., and Butler, C. (2015). Ability to be active: exploring children's active play in primary schools. *Int. J. Play* 4:2. doi: 10.1080/21594937.2015.1060569

72. Santangelo, L. (2012). Reducing stress in school-age girls through mindful yoga. *J. Pediatr. Health Care* 26, 45–56. doi: 10.1016/j.pedhc.2011.01.002

73. Sattelmair, J., and Ratey, J. (2009). Physically active play and cognition: an academic matter? *Am. J. Play* 1, 365–374.

74. Schaeffer, D. J., Krafft, C. E., Schwarz, N. F., Chi, L., Rodrigue, A. L., Pierce, J. E., et al. (2014). An 8-month exercise intervention alters frontotemporal white matter integrity in overweight children. *Psychophysiology* 51, 728–733. doi: 10.1111/psyp.12227

75. Schmidt, M., Jäger, K., Egger, F., Roebers, C. M., and Conzelmann, A. (2015). Cognitively engaging chronic physical activity, but not aerobic exercise, affects executive functions in primary school children: a

group-randomized controlled trial. *J. Sport Exerc. Psychol.* 37, 575–591. doi: 10.1123/jsep.2015-0069

76. Sibley, B. A., and Etnier, J. L. (2003). The relationship between physical activity and cognition in children: a meta-analysis. *Pediatr. Exerc. Sci.* 15, 243–256. doi: 10.1123/pes.15.3.243
77. Singh, A., Saliasi, E., van den Berg, V., Uijtdewilligen, L., de Groot, R. H. M., Jolles, J., et al. (2018). Effects of physical activity interventions on cognitive and academic performance in children and adolescents: a novel combination of a systematic review and recommendations from an expert panel. *Br. J. Sports Med.* doi: 10.1136/bjsports-2017-098136[Epub ahead of print].
78. Singh, A., Uijtdewilligen, L., Twisk, J. W. R., Van Mechelen, W., and Chinapaw, M. J. M. (2012). Physical activity and performance at school: a systematic review of the literature including a methodological quality assessment. *Arch. Pediatr. Adolesc. Med.* 166, 49–55. doi: 10.1001/archpediatrics.2011.716
79. Taimini, I. (2005). *The Science of Yoga.* Madras: The Theosophical Publishing House.
80. Telles, S., Hanumanthaiah, B. H., Nagarathna, R., and Nagendra, H. R. (1993). Improvement in static motor performance following yogic training of school children. *Percept. Mot. Skills* 76, 1264–1266. doi: 10.2466/pms.1993.76.3c.1264
81. Verburgh, L., Königs, M., Scherder, E. J., and Oosterlaan, J. (2014). Physical exercise and executive functions in preadolescent children, adolescents and young adults: a meta-analysis. *Br. J. Sports Med.* 48, 973–979. doi: 10.1136/bjsports-2012-091441
82. Verma, A., Uddhav, S., Ghanshyam Thakur, S., Devarao, D., Ranjit, K., and Bhogal, S. (2014). The effect of yoga practices on cognitive development in rural residential school children in India. *Natl. J. Lab. Med.* 3, 15–19.
83. World Medical Association (2013). Declaration of helsinki: ethical principles for medical research involving human subjects. *JAMA* 310, 2191–2194. doi: 10.1001/jama.2013.281053

Chapter 9

Increased Grey Matter Associated with LongTerm Sahaja Yoga Meditation: A Voxel-Based Morphometry Study

Sergio Elías Hernández[1], José Suero[2], Alfonso Barros[3], José Luis González-Mora[4], Katya Rubia[5]

[1] Department of Ingeniería Industrial, Universidad de La Laguna, Tenerife, Spain,

[2] Centro de Salud Jazmín, Sermas, Madrid, Spain,

[3] Department of Psychology, Universitat Jaume I, Castellón, Spain,

[4] Department of Fisiología, Universidad de La Laguna, Tenerife, Spain,

[5] Institute of Psychiatry, Psychology and Neuroscience, King's College London, London, United Kingdom

ABSTRACT

Objectives

To investigate regional differences in grey matter volume associated with the practice of Sahaja Yoga Meditation.

Citation: Increased Grey Matter Associated with Long-Term Sahaja Yoga Meditation: A Voxel-Based Morphometry Study. Hernández SE, Suero J, Barros A, González-Mora JL, Rubia K (2016). PLoS ONE 11(3): e0150757. doi:10.1371/journal. pone.0150757

Design

Twenty three experienced practitioners of Sahaja Yoga Meditation and twenty three non-meditators matched on age, gender and education level, were scanned using structural Magnetic Resonance Imaging and their grey matter volume were compared using Voxel-Based Morphometry.

Results

Grey matter volume was larger in meditators relative to non-meditators across the whole brain. In addition, grey matter volume was larger in several predominantly right hemispheric regions: in insula, ventromedial orbitofrontal cortex, inferior temporal and parietal cortices as well as in left ventrolateral prefrontal cortex and left insula. No areas with larger grey matter volume were found in non-meditators relative to meditators.

Conclusions

The study shows that long-term practice of Sahaja Yoga Meditation is associated with larger grey matter volume overall, and with regional enlargement in several right hemispheric cortical and subcortical brain regions that are associated with sustained attention, self-control, compassion and interoceptive perception. The increased grey matter volume in these attention and self-control mediating regions suggests use-dependent enlargement with regular practice of this meditation.

INTRODUCTION

The distribution of Grey Matter Volume (GMV) in our brain has an important role, among others on our mental health, our behavior and our cognitive functions. In normal development, GMV and cortical thickness increase in childhood, peak in adolescence and then decrease progressively with age across adulthood, due to a combination of progressive and regressive changes such as synaptic pruning, white matter development and neuronal loss in older age [1, 2]. There is considerable evidence of neuronal plasticity, which is higher in childhood but still considerable in adulthood with evidence that neuronal interconnections are modifiable networks that change in their structure as a consequence of behavioral or cognitive training or skill learning [3]. Thus, a close interrelationship has been demonstrated between localized use-dependent enlargement of GMV, of neuronal activation and the corresponding enhancement of functions or capabilities [4, 5]. Inversely, there is consistent evidence that abnormalities in GMV are associated with a reduction of cognitive and behavioral functions, reinforced by the fact that regional GMV is typically abnormal in mental disorders [6–8].

Research of meditation has experienced an important growth over the last two decades, among other things due to evidence for beneficial effects on mental and physical health and for therapeutic benefits for a range of mental and psychosomatic disorders [8–12].

Meditation has been associated with differences in brain structure both in cross-sectional imaging studies that compared long-term meditators with non-meditators and in short-term longitudinal imaging studies that assessed brain structure changes in novices after weeks of meditation practice (for review, see [13]). While longitudinal imaging studies can allow for more causal inferences, the cross-sectional comparison of long-term meditators with non-meditators can shed more light on the longer-term effects of meditation.

Several brain areas have consistently been shown to be different in several morphometry studies of long-term meditation [13], as well as been shown to be activated in fMRI studies of meditation, including the insular cortex [14–17], ventrolateral prefrontal cortex (VLPFC) [14, 18], somatomotor cortices [14, 19], inferior temporal gyrus [15, 18, 20, 21], fusiform gyrus [17, 18, 22], rostrolateral prefrontal cortex [14, 18, 22], anterior cingulate cortex [18, 23, 24] and hippocampus [25, 26]. Most of these areas mediate the top-down control or processing of attention or emotions, which is also the target of most meditation techniques [8, 13].

However, there is large heterogeneity among meditation techniques, which often involve very different practices and which are likely to have different effects on the brain. So far, the following meditation techniques have been studied using brain structure imaging techniques: Zen meditation [23, 27], Tibetan Buddhist meditation [22], Mindfulness meditation [25, 28, 29], Soham [30], Loving-kindness meditation [21], Brain wave vibration [18], or a combination of different meditation techniques within the same study: Chenrezig—Kriya—Shamatha—Vajrayana—Vipassana and Zazen [17, 20, 26, 31–33].

Meditation, as originally conceived in in the East by Patanjali [34, 35], is aimed to reduce thoughts to ultimately reach the state of thoughtless awareness which is considered a different state of consciousness where one is fully perceptually alert, yet has no thoughts. It is consequently described as a state of pure attention without any thought content [8]. This state of pure "contentless attention" appears to be subjectively associated with feelings of positive emotions and is described in Sanskrit as the state of "Sat Chit Ananda" which translates as a state of pure attention and joy [8, 36]. Sahaja Yoga Meditation (SYM) shares some goals with some other meditation techniques such as Mindfulness meditation, Loving-kindness meditation or other Buddhist meditations i.e., to be fully conscious on the present moment, to reduce the wandering mind and to increase compassion and love. However, one of the distinctive features of SYM is that the practitioner experiences the state of mental silence or "thoughtless awareness" on a regular basis in their meditation (if not necessarily daily), which makes it suitable to test the neural correlates of this unique state of thoughtless awareness which is the ultimate goal of traditional meditation, as originally conceived in the East [35].

SYM has been shown to have beneficial effects on a range of physical, mental and neurological disorders including: asthma [37], high blood pressure [38], menopause [39], epilepsy [40–42], depression [43], anxiety [38], work stress [44], and Attention-Deficit/Hyperactivity Disorder [45].

Several studies have used electroencephalography (EEG) to investigate the effects of SYM on brain activation. These studies showed increased theta and alpha activity over fronto-central parietal brain regions in long-term meditators compared to short-term meditators of SYM, which furthermore was associated with the state of mental silence. Furthermore, increase of theta activity over left frontal and limbic regions was associated with the feelings of joy associated with the state of mental silence. Fronto-parietal activation has been associated with sustained attention [46] while increased frontal alpha activity is thought to reflect a reduction in brain regions that mediate mental effort and external attention [47]. A more recent evoked potential study [48] showed that long-term SYM practice in long-term meditators compared to non-meditators was associated with increased efficiency of attention-related activity over predominantly right hemispheric fronto-central regions during the processing of salient emotional stimuli, which the authors interpreted as increased top-down control over fast automatic salience detection, presumably based on amygdala function. The findings reflect enhanced frontal lobe mediated top-down attention control and emotion regulation in long-term practitioners of SYM.

Only one previous study, from our group, used fMRI to assess brain activation during the meditation in the MRI scanner in long-term practitioners of SYM [49]. The study showed that, during their meditation, long-term SYM practitioners appear to pass through an initial intense neural self-control process necessary to silence their mind with neural activity in bilateral VLPFC/ insula and temporal regions. After this initial process, meditators experienced a relatively reduced brain activation concomitant with the deepening of the state of mental silence with a remaining activity in right rVLPFC/ insula and right middle/superior temporal cortex, with the activation in rVLPFC/insula being directly linked to the subjective depth of the meditation reported by the meditators. We interpreted the activity over rVLPFC as a reflection of the state of enhanced sustained attention during the state of mental silence.

In conclusion, functional imaging studies of long-term SYM practice report activation during meditation in fronto-parieto-temporal regions of sustained attention and in limbic regions of affect control [48–50]. However, nothing is known on the long-term effects of SYM on brain morphometry.

The aim of this study was therefore to test whether the acute activation differences we observed in fMRI during meditation relative to rest in long-term practitioners of SYM would also be reflected in regional grey matter differences on brain structure in long-term SYM practitioners relative to non-meditators, due to use-dependent plasticity. For this purpose, we compared GMV between a group of experienced SYM practitioners and a group of non-meditators, matched on age, education status and other demographic variables. Based on our fMRI and previous EEG findings on SYM [48–51], and based on the evidence for grey matter morphometry changes in long-term meditators of other meditation traditions in areas of affect and attention control [13], we hypothesised that experienced practitioners of SYM would have larger GMV in regions mediating attention and affect control, including: insula, VLPFC, ACC, the parieto-temporal junction, and temporo-limbic regions.

MATERIALS AND METHODS

Participants

Forty six right-handed, white Caucasian, healthy volunteers, between 21 and 63 years, participated. Twenty three experts in SYM (17 females) were compared with 23 non-meditators (17 females). Groups were matched on age, gender, education degree and body mass index (see Table 1). Volunteers had no physical or mental illness, no history of neurological disorders, and no addiction to nicotine, alcohol or drugs.

Table 1. Demographic characteristics of the group.

	Meditators Mean (SD)	Controls Mean (SD)	t(df = 44)	p-value*
Volunteers N°	23	23		
Age (years)	46.5 (11.4)	46.9 (10.9)	-0.13	0.89
Age range (years)	20.3–63.1	21.3–63.3		
Education degree, 0 to 6	3.78 (1.2)	4.04 (1.36)	0.69	0.50
Height (cm)	167.0 (8.8)	167.2 (7.6)	0.09	0.93
Weight (Kg)	69.5 (14.6)	71.7 (14.5)	0.53	0.60
Body mass index	24.9 (4.5)	25.5 (3.9)	0.54	0.60

* p-values represent group differences between meditators and controls using two-tailed independent samples t-tests.

Meditators were recruited from the local Tenerife SYM group in addition to SYM practitioners attending a seminar of Sahaja Yoga in Tenerife in January 2014. Controls were recruited through local and facebook advertisements. Controls did not practice any type of meditation or yoga. Only 3 controls reported a minimum meditation experience of less than 6 months practice.

Meditators had between 5 and 26 years of experience of daily meditation practice in SYM (mean:14.1 SD (6.1) years) and the average time dedicated daily to meditation per day was 84.7 (32.2) minutes. All participants filled in different questionnaires to evaluate their individual health status, education and age. Meditators additionally were given a questionnaire to register their experience in SYM such as: years of practice, total hours of meditation, average time dedicated to meditation per day and frequency of the perception of the state of mental silence. All participants signed informed consent to participate freely. This study was approved by the Ethics Committee of the University of La Laguna.

MRI Acquisition

All images were obtained on a 3T MRI Scanner, using an echo-planar-imaging gradient-echo sequence and an 8 channel head coil. A high-resolution T1-weighted three-dimensional inversion recovery spoiled gradient echo sequence was used to image the whole brain and the brainstem. A 3D fast spoiled-gradient-recalled pulse sequence was obtained with the following parameters: TR = 8.761 ms, TE = 1.736 ms, flip angle = 12°, matrix size = 256 x 256 pixels, spacing between slices and slice thickness = 1 mm, voxel resolution = 0.98 x 0.98 x 1 mm. Total acquisition time was 13 minutes.

Voxel-Based Morphometry

Voxel-based morphometry (VBM) [52] with DARTEL was conducted using SPM12 software package (Statistical Parametric Mapping software: http://www.fil.ion.ucl.ac.uk/spm/). Processing steps were performed as suggested by Ashburner [53]. VBM with DARTEL has been shown to be more sensitive than standard VBM [54] and provides results comparable to those achieved with manual segmentation [55].

The procedure followed these steps: 1. All T1-weighted anatomical images were displayed to screen to verify they were free from gross anatomical abnormalities. 2. For better registration, the T1 images were manually centred at the anterior commissure and reoriented according to the anterior–posterior commissure line. 3. Images were segmented into: Grey matter (GM), White matter (WM) and Cerebrum Spinal Fluid (CSF), using the New Segment procedure in SPM12, a segmentation that provides acceptable substitute for labor intensive manual estimates [56]. 4. The DARTEL routine inside SPM12 was used to spatially normalize the segmented images [54]. The image intensity of each voxel was modulated by the Jacobian determinants to ensure that regional differences in the total amount of GMV were conserved. 5. The registered images were then transformed to Montreal Neurological Institute (MNI) space using affine spatial normalization. 6. Finally, the normalized modulated GMV images were smoothed with a 4-mm full-width at half-maximum (FWHM) isotropic Gaussian kernel to increase the signal to noise ratio.

For each individual, total GM, WM and CSF were obtained with the Matlab script 'get_totals.m' [57] and used to calculate the individual Total Intracranial Volume (TIV) by summing the volumes of the three already mentioned components (GM, WM, CSF).

Statistical Analysis

Differences in GMV between meditators and controls were analyzed by conducting an ANCOVA analysis in SPM12. In order to control for inherent differences in brain structure, TIV, age, and gender were included as nuisance covariates [53]. An absolute threshold masking of 0.1 was used to avoid edge effects around the borders between GM and WM, which meant that voxels with a GM probability below 0.1 were removed from the analyses [53].

Because structural images display local variation in smoothness, cluster level correction was applied using random field theory and non-stationary correction [58]. Statistical thresholds were set at corrected p-values determined by non-stationary cluster-level correction with family wise error p(FWE-corr)<0.05 with uncorrected voxel-level of p<0.001.

To determine whether GMV clusters that differed between groups were correlated with the subjective measures of the meditation experience within the meditation group, a second-level SPM regression analysis was used to correlate these GMV clusters within the meditator group with the number of years of their meditation practice, the total hours meditated overall, the average time per day dedicated to meditation and the frequency of the perception of the state of mental silence. TIV, age, and gender were also included here as nuisance covariates.

RESULTS

Demographic Characteristics of the Groups

As shown in Table 1, the meditation and non-meditation groups did not differ in any demographic measures such as mean age, age range, height, weight, education status or body mass index.

The meditation group had been meditating between 5 and 26 years, with an average (standard deviation) of 14 (6) years. Their total hours of meditation were between 1213 to 12558 hours, with a mean of 5773 (3401) hours. Their daily practice was between 34 and 150 minutes, amounting to an average of 84 (32) minutes.

TIV and GMV Analysis

ANCOVA showed that TIV did not significantly differ between groups (F = 1.070, p = 0.307), using gender and age as nuisance covariates, meditators: TIV mean (SD) = 1385.3 mL (157.7), non-meditators: 1350.5 mL (172.4). In all other statistical analyses, TIV, gender and age were always entered as nuisance covariates.

ANCOVA with TIV, gender and age as nuisance covariates showed that overall whole brain GMV was significant different between groups (F = 10.445, p = 0.002), due to GMV being larger in meditators than non-meditators, meditators GMV mean (SD) = 653.4 mL (87.0); non-meditators 611.0 mL (74.6), (see Fig 1).

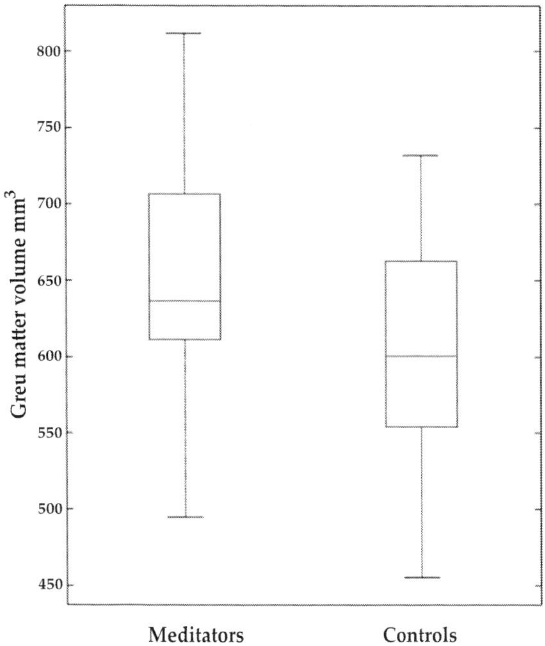

Figure 1. Box plot of grey matter volumes across the whole brain of meditators and controls. The central box represents the value from the lower to upper quartile (25th to 75th percentile). The middle line represents the median. The vertical line extends from the minimum to the maximum value.

Regional analyses showed that two brain regions were significantly larger in GMV in meditators compared to non-meditators (p (corrected) <0.05): the right anterior insula, extending rostrally into right ventromedial orbitofrontal cortex (vmOFC) and the right inferior temporal gyrus extended to the right fusiform gyrus. There was also a trend-finding for the right inferior parietal lobe (right angular gyrus), which was observed at a p-corrected cluster-level p = 0.069, (see Table 2 and Fig 2).

Table 2. Brain regions where meditators had significantly larger GMV than non-meditators.

Region	Side	Brodmann Area	Cluster size mm³	x	y	z	T of peak	Cluster p-value*
Insula, vmOFC	R	13, 47	564	30	10	-15	5,02	0,023
Inf. Temporal Gyrus, Fusiform Gyrus	R	20,37	739	52	-43	-21	4,43	0,037
Angular Gyrus**	R	39	476	52	-63	21	4,87	0,069**

x, y, z, (mm) MNI co-ordinates of peak,

* the corrected p-values were determined by non-stationary cluster-level correction based on family wise error p(FWE-corr)<0.05 with uncorrected voxel-level p<0.001, R = right side, Side = side of the hemisphere. vmOFC = ventromedial orbitofrontal cortex,

** only observed at a trend-level of p(FWE-corr) = 0.069.

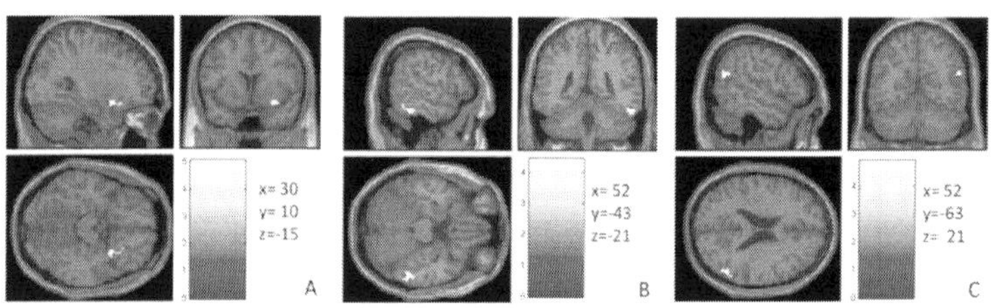

Figure 2. Areas where meditators exhibited larger GMV than non-meditators. (A) right insula/vmOFC, (B) right inferior temporal gyrus and (C) right angular gyrus. The color intensity represents *T*-statistic values at the voxel level.

Given that in our previous fMRI study we found that meditation elicits most prominently activation in VLPFC and insula [49], we wanted to test our a priori hypothesis that these regions would also be larger in the VBM analysis due to use-dependent plasticity. Therefore, to test our hypothesis of enhanced structure in VLPFC and insula, the data were re-analysed with a more lenient voxel-wise threshold of $p < 0.005$ and non-stationary cluster-level false discovery rate p(FDR-corr) $p < 0.05$. In fact, as expected, a cluster in left VLPFC (peak voxel MNI coordinates x, y, z: -37, 49, -13: cluster p-corr = 0.04) and a cluster in left anterior insula (peak voxel MNI coordinates x, y, z: -28, 10, -9, cluster p-corr = 0.04) were larger in meditators relative to non-meditators, (see Fig 3). There were no regions where non-meditators had larger GMV than mediators.

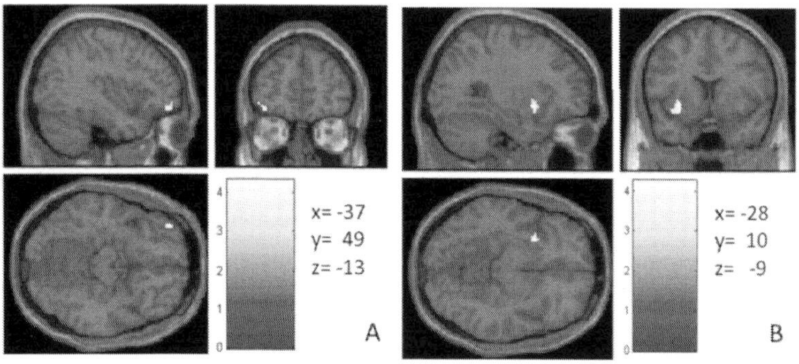

Figure 3. A priori hypothesised regions that showed larger GMV in meditators relative to controls at a more lenient threshold. (A) left VLPFC (B) left anterior insula. The color intensity represents *T*-statistic values at the voxel level.

Correlation Analyses

There were no significant correlations within the meditation group between the total GMV or the GMV in any of the regions that differed between groups and the meditators' experience, including number of years of their meditation practice, the total hours meditated overall, the average time per day dedicated to meditation or the frequency of the perception of the state of mental silence.

DISCUSSION

To our knowledge, this is the first study that tested the association between long-term SYM, which elicits the state of mental silence on a regular basis, and brain morphometry. Long-term practitioners of SYM showed larger GMV in predominantly right hemispheric brain areas such as the ventromedial orbitofrontal cortex (vmOFC), the insula, and the inferior temporal lobe. These areas are associated with affect and attention control, interoception and autonomic function surveillance. At a more lenient level, we also observed enhanced GMV in left VLPFC and left insula. The findings extend our previous fMRI results by showing that attention and limbic regions that are activated during the practice of meditation in long-term meditators such as VLPFC/insula and temporal lobe are also larger in structure in long-term meditators, presumably due to use-dependent plastic effects of years of SYM practice on these brain structures.

The findings of larger right and left insula in practitioners of SYM relative to non-meditators are in line with the effects observed in other meditation techniques. The anterior insular cortex is one of the most replicated areas that have been shown to be enlarged in structure in studies of meditation, most commonly in the right hemisphere [14, 15, 18], but also bilaterally as in this study [17]. The insula is associated with interoception, visceral and emotional self-awareness, including awareness of respiration and heart rate [59]. The anterior insula as well as inferior temporal regions are also part of the cingular-opercular salience detection network and the insula has been shown to mediate tonic and intrinsic alertness [60], which may reflect the subjective experience during meditation of enhanced alertness and attention during the state of mental silence [61]. This difference between meditators and controls presumably reflects the inner attention training during SYM that is common with other types of meditation like Vipassana or Mindfulness Meditation [62, 63] and which may be associated with enhanced body-self-awareness and increased intrinsic self-awareness and alertness during the meditation [13, 60].

This larger GMV in the insula may also play a role in the emotional and clinical benefits shown to be engendered by meditation [9, 12], given that larger GMV in bilateral insula has been associated with "good life", traits concerned with personal growth, self-acceptance, purpose in life and autonomy (among others), and been shown to be a good predictor of good health [64]. Right insula GMV increase has also been associated with improvements to identify and describe emotions in the self after a brief 8 weeks- Mindfulness Based Stress Reduction training program [65]. Emotional intelligence has also been shown to correlate positively with larger GMV in left insula and rVLPFC in a study in 328 university students [66].

The right insula cluster extended to the vmOFC that is well connected with sensory regions and part of the limbic affect regulation system, exerting top-down control over amygdala, ventral striatum and hypothalamus [67]. While amygdala and striatum store and generate a more automatic reaction to stimuli based on past experiences, the vmOFC generates a more flexible assessment to counterbalance the automatic reactivity in amygdala and striatum and has hence an important role in top-down emotional regulation and reappraisal of negative emotional states [48, 68, 69].

The enlarged GMV in this vmOFC affect control region may be related to the subjective experience of the state of mental silence, which is associated with feelings of positive emotions and a detached "witness state". In EEG studies, the subjective feelings of happiness during SYM have been associated with activation over fronto-temporal regions [70]. Furthermore, long-term practitioners of SYM showed greater emotional stability and greater detachment than non-meditators during the exposure to stressful aversive film viewing, as shown in lower subjective emotional ratings of the response to the stressful stimuli, a reduction in parasympathetic stress indicators such as skin conductance measures, and reduced emotional arousal, indexed by alpha and gamma EEG oscillations [51]. Last, long-term SYM practitioners relative to non-meditators showed increased attention-related right hemispheric fronto-central activation during the processing of emotionally salient stimuli, suggesting increased frontal top-down control over automatic saliency processing regions and hence enhanced cognitive and affective control [48]. This enhanced affect control, in particular over negative emotions, is likely to be mediated by vmOFC top-down control over limbic brain regions. The enlarged GMV in vmOFC could also potentially be associated with findings that several weeks of SYM treatment has been shown to reduce symptoms of depression and anxiety, which one could speculate may reflect an effect on ventromedial fronto-limbic brain systems of affect regulation [38, 43, 44].

We found a more dorsal cluster of rVLPFC and right insula to be activated in fMRI in long-term SYM mediators during their meditation in the scanner which was furthermore directly associated with the subjective depth of the state of mental silence [49]. Although the peak of the frontal cluster was in a more dorsal location, some of the cluster extent overlapped with the here observed ventromedial prefrontal location. Right fronto-parieto-temporal activation was also correlated with the state of mental silence in an EEG study of SYM [47, 48]. We hypothesized that the more dorsal location in rVLPFC that was associated with the "acute meditation" was related to the inhibitory self-control process needed to reject unwanted thoughts or distractions, which is crucial to reach and sustain the state of mental silence [49] and which is typically mediated by rVLPFC [71, 72]. Ventrolateral prefrontal cortex, anterior insula, inferior parietal and inferior temporal lobe together play a crucial role in attention, as they form part of the ventral attention system [73–75]. The practice of meditation has been considered an effective attention training [8], illustrated by improved performance in cognitive attention tasks in long-term meditators [12, 63, 76, 77] and this may well have led to plastic changes in regions of the right hemispheric ventral attention system that mediates sustained attention.

The enlarged GMV in right inferior temporal lobe is intriguing. As mentioned above, the bilateral inferior posterior temporal lobes are part of a generic cognitive control network that is activated during higher level and attention demanding tasks, independent of the task in question, and thus part of the dorsal and ventral attention systems [78–80]. The regular practice of meditation, if considered an effective attention and cognitive self-control training, may lead to the enlargement of some parts of this generic attention network. Previous morphometry studies of meditation have found larger GMV in inferior temporal gyrus, albeit in the left hemisphere [15, 18, 20, 21]. The integrity of this region is also key to prosopagnosia, where the emotional familiarity of a face is no longer recognised [81, 82]. Alternatively, the right inferior temporal lobe has also been associated with socio-emotional competence. Thus, a recent study found that right inferior temporal GMV was positively associated with individual variations of gratitude so that individuals with larger GMV expressed more gratitude [83]. Similarly, the increased volume of this region was associated in healthy individuals with higher competence in interpreting the intentions of others [84]. Also, WM volume in a slightly more superior right posterior temporal region has been associated with autism spectrum traits in healthy participants [85]. It thus appears that the integrity of this region may be related to socio-emotional abilities.

The right angular gyrus cluster that was larger at a trend-level (p-corr = 0.69) in long-term SYM practitioners is also part of the right temporo-parietal junction that is not only a crucial part of posterior attention networks [73–75] but has also been related to empathy. Thus the temporo-parietal junction is activated when we try to understand the feelings and thought of others or show empathy for others [86]. These are important capacities for an enriched emotional and social life that facilitates the emotions of love and compassion [87]. Altruism has also been shown to be linearly correlated with GMV in the temporo-parietal junction in a VBM research [88].

Although we observed at a more lenient threshold activation in left insula and left VLPFC, the areas that showed most significantly larger GMV in long-term meditators were in the right hemisphere. The findings are in line with the "acute" changes we observed in long-term meditators of SYM during their meditation in our fMRI study, where initially bilateral VLPFC/insula and temporal regions were activated which then became more prominently right hemispheric with the progressive depth of meditation. Predominantly right-hemispheric fronto-temporal activation [49] and fronto-temporo-parietal activation were also associated with the depth of thoughtless awareness in EEG studies [47]. Attention systems are more prominently right-hemispheric [73] which would fit the notion that meditation is a powerful attention training that enhances the plasticity of attention and saliency processing in right hemispheric fronto-insular-parieto-temporal regions. Furthermore, the right hemisphere is also more closely connected to the limbic system [89, 90] and mediates interoceptive functions [59, 91].

In addition to larger GMV in these brain regions of interoceptive perception, attention and/or affect processing, we also found that long-term practitioners of SYM compared to non-meditators had overall significantly larger GMV across the whole brain. This overall larger GMV, not previously observed in other meditation techniques, may be related to the differences of SYM relative to other meditation practices, most prominently the frequent achievement of the state of mental silence which is subjectively experienced as an altered state of consciousness. Given that GMV decreases with age and this is associated with the normal aging process [2], one could potentially speculate that long-term meditation is associated with a delay of normal age-related decline in GMV and hence a younger brain structure pattern overall.

An important limitation of this study, common to all cross-sectional studies of brain morphometry, is that brain morphometry differences cannot be attributed solely to meditation practice. Although we matched groups by age, gender, education level, and body mass, we cannot exclude that differences in lifestyles, or other unmeasured factors that may have differed between groups previous to the practice of meditation, could have influenced the brain morphometry measures.

Causal inferences are further hampered by the fact that we did not find any correlations between the brain structure changes and meditation–related measures such as years or hours of meditation in the SYM group, time dedicated to meditation per day or the subjective depth of the meditation experience overall. It is possible that the lack of correlation suggests a ceiling effect that may precede the 5 years of minimum accumulated practice of meditation in this study. Previous morphometric studies of meditation have shown inconsistent associations between GMV and the meditators' experience (mainly years practising meditation or total hours meditated) [13–15, 20]. Given the lack of correlations, we cannot exclude that brain changes in long-term meditators may be associated with some other mediating or moderating factors that differ between people who chose to meditate and those who don't. Future longitudinal VBM studies of SYM may be more powerful to infer direct causal effects between meditation practice and brain structure changes.

CONCLUSIONS

This is the first brain morphometry characterization of the long-term practice of SYM. We show that long-term practitioners of SYM compared with non-meditators have statistically significant larger GMV across the whole brain, not previously observed in practitioners of any other meditation technique. Furthermore, GMV differed in several predominantly right hemispheric regions: in insula, ventromedial orbitofrontal cortex, inferior temporal and parietal cortices as well as in left ventrolateral prefrontal cortex and left insula.

These areas have been shown to be associated with sustained attention/cognitive control, emotion control, interoceptive perception and feelings of compassion, suggesting that long-term SYM practice may potentially enhance the functions mediated by these regions and consequently lead to neuroplastic enlargements.

ACKNOWLEDGMENTS

We acknowledge the support of MRI services for Biomedical Studies (Servicio de Resonancia Magnética para Investigaciones Biomédicas) of the University of La Laguna. We warmly thank all the volunteers for their participation in this study.

AUTHOR CONTRIBUTIONS

Conceived and designed the experiments: SEH JS AB JLGM KR. Performed the experiments: SEH KR. Analyzed the data: SEH AB KR. Contributed reagents/materials/analysis tools: SEH. Wrote the paper: SEH JS AB JLGM KR.

REFERENCES

1. Giedd JN, Raznahan A, Alexander-Bloch A, Schmitt E, Gogtay N, Rapoport JL. Child Psychiatry Branch of the National Institute of Mental Health Longitudinal Structural Magnetic Resonance Imaging Study of Human Brain Development. Neuropsychopharmacology. 2015;40(1):43–9. WOS:000346146400005. pmid:25195638
2. Mills KL, Tamnes CK. Methods and considerations for longitudinal structural brain imaging analysis across development. Developmental Cognitive Neuroscience. 2014;9:172–90. WOS:000337990300015. pmid:24879112
3. Schlaug G. Musicians and music making as a model for the study of brain plasticity. Progress in brain research. 2015;217:37–55. MEDLINE:25725909. pmid:25725909
4. Taubert M, Villringer A, Ragert P. Learning-Related Gray and White Matter Changes in Humans: An Update. Neuroscientist. 2012;18(4):320–5. WOS:000306732400007. pmid:22013150
5. Zatorre RJ, Fields RD, Johansen-Berg H. Plasticity in gray and white: neuroimaging changes in brain structure during learning. Nature Neuroscience. 2012;15(4):528–36. WOS:000302114500010. pmid:22426254
6. Grieve SM, Korgaonkar MS, Koslow SH, Gordon E, Williams LM. Widespread reductions in gray matter volume in depression. Neuroimage-Clinical. 2013;3:332–9. WOS:000209276900036. pmid:24273717
7. Singh S, Modi S, Goyal S, Kaur P, Singh N, Bhatia T, et al. Functional and structural abnormalities associated with empathy in patients with schizophrenia: An fMRI and VBM study. Journal of Biosciences. 2015;40(2):355–64. WOS:000355068100015. pmid:25963262
8. Rubia K. The neurobiology of Meditation and its clinical effectiveness in psychiatric disorders. Biological Psychology. 2009;82(1):1–11. WOS:000269112300001. pmid:19393712
9. Sedlmeier P, Eberth J, Schwarz M, Zimmermann D, Haarig F, Jaeger S, et al. The Psychological Effects of Meditation: A Meta-Analysis. Psychological Bulletin. 2012;138(6):1139–71. WOS:000310116600006. pmid:22582738
10. Khoury B, Lecomte T, Fortin G, Masse M, Therien P, Bouchard V, et al. Mindfulness-based therapy: A comprehensive meta-analysis. Clinical

Psychology Review. 2013;33(6):763–71. WOS:000322690200004. pmid:23796855

11. Chen KW, Berger CC, Manheimer E, Forde D, Magidson J, Dachman L, et al. Meditative Therapies for Reducing Anxiety: A Systematic Review and Meta-Analysis of Randomized Controlled Trials. Depression and Anxiety. 2012;29(7):545–62. WOS:000305936200002. pmid:22700446
12. Chiesa A, Serretti A. A systematic review of neurobiological and clinical features of mindfulness meditations. Psychological Medicine. 2010;40(8):1239–52. WOS:000280158500002. pmid:19941676
13. Fox KCR, Nijeboer S, Dixon ML, Floman JL, Ellamil M, Rumak SP, et al. Is meditation associated with altered brain structure? A systematic review and meta-analysis of morphometric neuroimaging in meditation practitioners. Neuroscience and Biobehavioral Reviews. 2014;43:48–73. WOS:000338607400004. pmid:24705269
14. Lazar SW, Kerr CE, Wasserman RH, Gray JR, Greve DN, Treadway MT, et al. Meditation experience is associated with increased cortical thickness. Neuroreport. 2005;16(17):1893–7. WOS:000233546200005. pmid:16272874
15. Hoelzel BK, Ott U, Gard T, Hempel H, Weygandt M, Morgen K, et al. Investigation of mindfulness meditation practitioners with voxel-based morphometry. Social Cognitive and Affective Neuroscience. 2008;3(1):55–61. WOS:000256174000007. pmid:19015095
16. Tang Y-Y. Mechanism of Integrative Body-Mind Training. Neuroscience Bulletin. 2011;27(6):383–8. WOS:000297726300003. pmid:22108815
17. Luders E, Kurth F, Mayer EA, Toga AW, Narr KL, Gaser C. The unique brainan atomy of meditation practitioners: alterations in cortical gyrification. Frontiers in Human Neuroscience. 2012;6. 34 WOS:000301122500001. pmid:22393318
18. Kang D-H, Jo HJ, Jung WH, Kim SH, Jung Y-H, Choi C-H, et al. The effect of meditation on brain structure: cortical thickness mapping and diffusion tensor imaging. Social Cognitive and Affective Neuroscience. 2013;8(1):27–33. WOS:000313649700004. pmid:22569185
19. Grant JA, Courtemanche J, Rainville P. A non-elaborative mental stance and decoupling of executive and pain-related cortices predicts low pain sensitivity in Zen meditators. Pain. 2011;152(1):150–6. WOS:000285410800025. pmid:21055874

20. Luders E, Toga AW, Lepore N, Gaser C. The underlying anatomical correlates of long-term meditation: Larger hippocampal and frontal volumes of gray matter. Neuroimage. 2009;45(3):672–8. WOS:000264378400004. pmid:19280691
21. Leung M-K, Chan CCH, Yin J, Lee C-F, So K-F, Lee TMC. Increased gray matter volume in the right angular and posterior parahippocampal gyri in loving-kindness meditators. Social Cognitive and Affective Neuroscience. 2013;8(1):34–9. WOS:000313649700005. pmid:22814662
22. Vestergaard-Poulsen P, van Beek M, Skewes J, Bjarkam CR, Stubberup M, Bertelsen J, et al. Long-term meditation is associated with increased gray matter density in the brain stem. Neuroreport. 2009;20(2):170–4. WOS:000263318900014. pmid:19104459
23. Grant JA, Courtemanche J, Duerden EG, Duncan GH, Rainville P. Cortical Thickness and Pain Sensitivity in Zen Meditators. Emotion. 2010;10(1):43–53. WOS:000274400800007. pmid:20141301
24. Tang Y-Y, Lu Q, Geng X, Stein EA, Yang Y, Posner MI. Short-term meditation induces white matter changes in the anterior cingulate. Proceedings of the National Academy of Sciences of the United States of America. 2010;107(35):15649–52. WOS:000281468500064. pmid:20713717
25. Hoelzel BK, Carmody J, Vangel M, Congleton C, Yerramsetti SM, Gard T, et al. Mindfulness practice leads to increases in regional brain gray matter density. Psychiatry Research-Neuroimaging. 2011;191(1):36–43. WOS:000288591100006.
26. Luders E, Kurth F, Toga AW, Narr KL, Gaser C. Meditation effects within the hippocampal complex revealed by voxel-based morphometry and cytoarchitectonic probabilistic mapping. Frontiers in Psychology. 2013;4. WOS:000331106500001.
27. Pagnoni G, Cekic M. Age effects on gray matter volume and attentional performance in Zen meditation. Neurobiology of Aging. 2007;28(10):1623–7. WOS:000249062800016. pmid:17655980
28. Farb NAS, Segal ZV, Anderson AK. Mindfulness meditation training alters cortical representations of interoceptive attention. Social cognitive and affective neuroscience. 2013;8(1):15–26. MEDLINE:22689216. pmid:22689216

29. Singleton O, Hoelzel BK, Vangel M, Brach N, Carmody J, Lazar SW. Change in brainstem gray matter concentration following a mindfulness-based intervention is correlated with improvement in psychological well-being. Frontiers in Human Neuroscience. 2014;8. WOS:000332081700001.

30. Kumar U, Guleria A, Kishan SSK, Khetrapal CL. Effect of SOHAM Meditation on Human Brain: A Voxel-Based Morphometry Study. Journal of Neuroimaging. 2014;24(2):187–90. WOS:000332095700014. pmid:23773541

31. Luders E, Kurth F, Thompson PM, Hong JY, Gutman B, Phillips OR, et al. The unique anatomy of the hippocampus in meditation practitioners. European Journal of Neurology. 2012;19:76-. WOS:000309359700144.

32. Luders E. Exploring age-related brain degeneration in meditation practitioners. In: Sequeira S, editor. Advances in Meditation Research: Neuroscience and Clinical Applications. Annals of the New York Academy of Sciences. 13072014. p. 82–8.

33. Luders E, Cherbuin N, Kurth F. Forever Young(er): potential age-defying effects of long-term meditation on gray matter atrophy. Frontiers in Psychology. 2015;5. WOS:000348190800001.

34. Bourgeault C. The 'Yoga Sutras' of Patanjali. Parabola-Myth Tradition and the Search for Meaning. 2002;27(2):92-+. WOS:000180508600026.

35. Kokodoko A. The Yoga Sutra of Patanjali. Library Journal. 2014;139(6):96-. WOS:000333551300146.

36. Manocha R. Meditation, mindfulness and mind-emptiness. Acta Neuropsychiatrica. 2011;23(1):46–7. WOS:000285971500011.

37. Manocha R, Marks GB, Kenchington P, Peters D, Salome CM. Sahaia yoga in the management of moderate to severe asthma: a randomised controlled trial. Thorax. 2002;57(2):110–5. WOS:000173769800005. pmid:11828038

38. Chung S-C, Brooks MM, Rai M, Balk JL, Rai S. Effect of Sahaja Yoga Meditation on Quality of Life, Anxiety, and Blood Pressure Control. Journal of Alternative and Complementary Medicine. 2012;18(6). WOS:000306356200012.

39. Manocha R, Semmar B, Black D. A pilot study of a mental silence form of meditation for women in perimenopause. Journal of Clinical Psychology in Medical Settings. 2007;14(3):266–73. WOS:000249818400010.

40. Panjwani U, Gupta HL, Singh SH, Selvamurthy W, Rai UC. Effect of Sahaja yoga practice on stress management in patients of epilepsy. Indian journal of physiology and pharmacology. 1995;39(2):111–6. MEDLINE:7649596. pmid:7649596
41. Panjwani U, Selvamurthy W, Singh SH, Gupta HL, Thakur L, Rai UC. Effect of Sahaja yoga practice on seizure control and EEG changes in patients of epilepsy. Indian Journal of Medical Research. 1996;103:165–72. WOS:A1996UG10500009. pmid:9062044
42. Panjwani U, Selvamurthy W, Singh SH, Gupta HL, Mukhopadhyay S, Thakur L. Effect of Sahaja yoga meditation on Auditory Evoked Potentials (AEP) and Visual Contrast Sensitivity (VCS) in epileptics. Applied Psychophysiology and Biofeedback. 2000;25(1):1–12. WOS:000087119500001. pmid:10832506
43. Morgan A. Sahaja Yoga: an ancient path to modern mental health? Transpersonal Psychology Review ed: Transpersonal Psychology; 2001.
44. Manocha R, Black D, Sarris J, Stough C. A Randomized, Controlled Trial of Meditation for Work Stress, Anxiety and Depressed Mood in Full-Time Workers. Evidence-Based Complementary and Alternative Medicine. 2011:1–8. WOS:000293644300001.
45. Harrison L, Manosh R, Rubia K. Sahaja Yoga Meditation as a family treatment program for attention deficit hyperactivity disorder children. Journal of Clinical Psychology and Psychiatry. 2004;9(4):479–97.
46. Sauseng P, Hoppe J, Klimesch W, Gerloff C, Hummel FC. Dissociation of sustained attention from central executive functions: local activity and interregional connectivity in the theta range. European Journal of Neuroscience. 2007;25(2):587–93. WOS:000244783000029. pmid:17284201
47. Aftanas LI, Golocheikine SA. Human anterior and frontal midline theta and lower alpha reflect emotionally positive state and internalized attention: high-resolution EEG investigation of meditation. Neuroscience Letters. 2001;310(1):57–60. WOS:000170855900015. pmid:11524157
48. Reva NV, Pavlov SV, Loktev KV, Korenyok VV, Aftanas LI. Influence of Long-Term Sahaja Yoga Meditation Practice on Emotional Processing in the Brain: An Erp Study. Neuroscience. 2014;281:195–201. WOS:000343860700017.

49. Hernandez SE, Suero J, Rubia K, Gonzalez-Mora JL. Monitoring the Neural Activity of the State of Mental Silence While Practicing Sahaja Yoga Meditation. Journal of Alternative and Complementary Medicine. 2015;21(3):175–9. WOS:000350749200007. pmid:25671603
50. Aftanas LI, Golocheikine SA. Non-linear dynamic complexity of the human EEG during meditation. Neuroscience Letters. 2002;330(2):143–6. Pii s0304-3940(02)00745-0 WOS:000178725400005. pmid:12231432
51. Aftanas L, Golosheykin S. Impact of regular meditation practice on EEG activity at rest and during evoked negative emotions. International Journal of Neuroscience. 2005;115(6):893–909. WOS:000229439400012. pmid:16019582
52. Ashburner J, Friston KJ. Voxel-based morphometry—The methods. Neuroimage. 2000;11(6):805–21. WOS:000087963600018. pmid:10860804
53. VBM Tutorial. Available: [Internet]. 2010. Available: http://www. fil. ion. ucl. ac. uk/~john/misc/VBMclass10. pdf.
54. Ashburner J. A fast diffeomorphic image registration algorithm. Neuroimage. 2007;38(1):95–113. WOS:000249992900011. pmid:17761438
55. Bergouignan L, Chupin M, Czechowska Y, Kinkingnehun S, Lemogne C, Le Bastard G, et al. Can voxel based morphometry, manual segmentation and automated segmentation equally detect hippocampal volume differences in acute depression? Neuroimage. 2009;45(1):29–37. WOS:000263862900004. pmid:19071222
56. Malone IB, Leung KK, Clegg S, Barnes J, Whitwell JL, Ashburner J, et al. Accurate automatic estimation of total intracranial volume: A nuisance variable with less nuisance. Neuroimage. 2015;104:366–72. WOS:000345393800035. pmid:25255942
57. Ridgway. Get_totals. m Matlab script
58. Hayasaka S, Phan KL, Liberzon I, Worsley KJ, Nichols TE. Nonstationary cluster-size inference with random field and permutation methods. Neuroimage. 2004;22(2):676–87. WOS:000222091600020. pmid:15193596
59. Craig AD. Emotional moments across time: a possible neural basis for time perception in the anterior insula. Philosophical Transactions of the Royal Society B-Biological Sciences. 2009;364(1525):1933–42. WOS:000267281300012.

60. Sadaghiani S, D'Esposito M. Functional Characterization of the Cingulo-Opercular Network in the Maintenance of Tonic Alertness. Cerebral Cortex. 2015;25(9):2763–73. WOS:000361464000037. pmid:24770711

61. Clemens B, Zvyagintsev M, Sack A, Heinecke A, Willmes K, Sturm W. Revealing the Functional Neuroanatomy of Intrinsic Alertness Using fMRI: Methodological Peculiarities. Plos One. 2011;6(9). WOS:000295941300033.

62. Hoelzel BK, Ott U, Hempel H, Hackl A, Wolf K, Stark R, et al. Differential engagement of anterior cingulate and adjacent medial frontal cortex in adept meditators and non-meditators. Neuroscience Letters. 2007;421(1):16–21. WOS:000248152500004. pmid:17548160

63. Lutz A, Slagter HA, Dunne JD, Davidson RJ. Cognitive-emotional interactions—Attention regulation and monitoring in meditation. Trends in Cognitive Sciences. 2008;12(4):163–9.

64. Lewis GJ, Kanai R, Rees G, Bates TC. Neural correlates of the 'good life': eudaimonic well-being is associated with insular cortex volume. Social Cognitive and Affective Neuroscience. 2014;9(5):615–8. WOS:000336489000007. pmid:23512932

65. Santarnecchi E, D'Arista S, Egiziano E, Gardi C, Petrosino R, Vatti G, et al. Interaction between Neuroanatomical and Psychological Changes after Mindfulness-Based Training. Plos One. 2014;9(10). WOS:000343942100010.

66. Tan Y, Zhang Q, Li W, Wei D, Qiao L, Qiu J, et al. The correlation between Emotional Intelligence and gray matter volume in university students. Brain and Cognition. 2014;91:100–7. WOS:000345487400013. pmid:25282329

67. Pessoa L. Emotion and cognition and the amygdala From "what is it?" to "what's to be done?". Neuropsychologia. 2010;48(12):3416–29. WOS:000284017300005. pmid:20619280

68. Goldin PR, Gross JJ. Effects of Mindfulness-Based Stress Reduction (MBSR) on Emotion Regulation in Social Anxiety Disorder. Emotion. 2010;10(1):83–91. WOS:000274400800011. pmid:20141305

69. Schoenbaum G, Takahashi Y, Liu T-L, McDannald MA. Does the orbitofrontal cortex signal value? Critical Contributions of the Orbitofrontal Cortex to Behavior. 2011;1239:87–99. WOS:000299228800009.

70. Aftanas LI, Golocheikine SA. Linear and non-linear concomitants of altered state of consciousness during meditation: High resolution EEG investigation. International Journal of Psychophysiology. 2002;45(1–2):158-. WOS:000177095500403.
71. Rubia K, Smith AB, Brammer MJ, Taylor E. Right inferior prefrontal cortex mediates response inhibition while mesial prefrontal cortex is responsible for error detection. Neuroimage. 2003;20(1):351–8. WOS:000185746400031. pmid:14527595
72. Aron AR, Fletcher PC, Bullmore ET, Sahakian BJ, Robbins TW. Stop-signal inhibition disrupted by damage to right inferior frontal gyrus in humans. Nature Neuroscience. 2003;6(2):115–6. WOS:000180669100009. pmid:12536210
73. Corbetta M, Kincade JM, Shulman GL. Neural systems for visual orienting and their relationships to spatial working memory. Journal of Cognitive Neuroscience. 2002;14(3):508–23. WOS:000174711800016. pmid:11970810
74. Patel GH, Yang D, Jamerson EC, Snyder LH, Corbetta M, Ferrera VP. Functional evolution of new and expanded attention networks in humans (vol 112, pg 9454, 2015). Proceedings of the National Academy of Sciences of the United States of America. 2015;112(38):E5377–E. WOS:000361525100023.
75. Thakral PP. The neural substrates associated with inattentional blindness. Consciousness and Cognition. 2011;20(4):1768–75. WOS:000297278100075. pmid:21481608
76. Srinivasan N, Baijal S. Concentrative meditation enhances preattentive processing: a mismatch negativity study. Neuroreport. 2007;18(16):1709–12. WOS:000250329300015. pmid:17921873
77. Lutz A, Greischar LL, Perlman DM, Davidson RJ. BOLD signal in insula is differentially related to cardiac function during compassion meditation in experts vs. novices. Neuroimage. 2009;47(3):1038–46. WOS:000268926200031. pmid:19426817
78. Hugdahl K, Raichle ME, Mitra A, Specht K. On the existence of a generalized non-specific task-dependent network. Frontiers in Human Neuroscience. 2015;9. WOS:000360267400001.
79. Niendam TA, Laird AR, Ray KL, Dean YM, Glahn DC, Carter CS. Meta-analytic evidence for a superordinate cognitive control network subserving diverse executive functions. Cognitive Affective & Behavioral Neuroscience. 2012;12(2):241–68. WOS:000303476100001.

80. Fox MD, Zhang D, Snyder AZ, Raichle ME. The Global Signal and Observed Anticorrelated Resting State Brain Networks. Journal of Neurophysiology. 2009;101(6):3270–83. WOS:000266398500048. pmid:19339462

81. Barton JJS. Structure and function in acquired prosopagnosia: Lessons from a series of 10 patients with brain damage. Journal of Neuropsychology. 2008;2:197–225. WOS:000270259600012. pmid:19334311

82. Dinkelacker V, Grueter M, Klaver P, Grueter T, Specht K, Weis S, et al. Congenital prosopagnosia: multistage anatomical and functional deficits in face processing circuitry. Journal of Neurology. 2011;258(5):770–82. WOS:000290453400007. pmid:21120515

83. Zahn R, Garrido G, Moll J, Grafman J. Individual differences in posterior cortical volume correlate with proneness to pride and gratitude. Social Cognitive and Affective Neuroscience. 2014;9(11):1676–83. WOS:000345846800005. pmid:24106333

84. Lewis PA, Rezaie R, Brown R, Roberts N, Dunbar RIM. Ventromedial prefrontal volume predicts understanding of others and social network size. Neuroimage. 2011;57(4):1624–9. WOS:000293054100036. pmid:21616156

85. von dem Hagen EAH, Nummenmaa L, Yu R, Engell AD, Ewbank MP, Calder AJ. Autism Spectrum Traits in the Typical Population Predict Structure and Function in the Posterior Superior Temporal Sulcus. Cerebral Cortex. 2011;21(3):493–500. WOS:000287488100001. pmid:20439317

86. Saxe R, Wexler A. Making sense of another mind: The role of the right temporo-parietal junction. Neuropsychologia. 2005;43(10):1391–9. WOS:000230809100001. pmid:15936784

87. Saxe R, Kanwisher N. People thinking about thinking people—The role of the temporo-parietal junction in "theory of mind". Neuroimage. 2003;19(4):1835–42. WOS:000185079000053. pmid:12948738

88. Morishima Y, Schunk D, Bruhin A, Ruff CC, Fehr E. Linking Brain Structure and Activation in Temporoparietal Junction to Explain the Neurobiology of Human Altruism. Neuron. 2012;75(1):73–9. WOS:000306539600010. pmid:22794262

89. Rotenberg VS, Weinberg I. Human memory, cerebral hemispheres, and the limbic system: A new approach. Genetic Social and General Psychology Monographs. 1999;125(1):45–70. WOS:000078515600003.

90. Borod JC, Bloom RL, Brickman AM, Nakhutina L, Curko EA. Emotional processing deficits in individuals with unilateral brain damage. Applied neuropsychology. 2002;9(1):23–36. MEDLINE:12173747. pmid:12173747
91. Pollatos O, Gramann K, Schandry R. Neural systems connecting interoceptive awareness and feelings. Human Brain Mapping. 2007;28(1):9–18. WOS:000242944100002. pmid:16729289

Chapter 10

Yoga Practice Reduces the Psychological Distress Levels of Prison Inmates

Anis Sfendla[1] , Petter Malmström[2] , Sara Torstensson[2] and Nóra Kerekes [2]

[1] Department of Biology, Faculty of Sciences, Abdelmalek Essaâdi University, Tetouan, Morocco

[2] Department of Health Sciences, University West, Trollhättan, Sweden

ABSTRACT

Background

Psychiatric ill-health is prevalent among prison inmates and often hampers their rehabilitation. Rehabilitation is crucial for reducing recidivistic offending. A few studies have presented evidence of the positive effect of yoga on the well-being of prison inmates. The conclusion of those previous studies that yoga is an effective method in the rehabilitation process of inmates, and deserves and requires further attention.

Citation: Yoga Practice Reduces the Psychological Distress Levels of Prison Inmates. Anis Sfendla, Petter Malmström, Sara Torstensson and Nóra Kerekes. Front. Psychiatry 9:407. doi: 10.3389/fpsyt.2018.00407.

Aims

The current study aimed to evaluate the effect of 10 weeks of yoga practice on the mental health profile, operationalized in the form of psychological distress, of inmates.

Methods

One hundred and fifty-two volunteer participants (133 men; 19 women) were randomly placed in either of two groups: to participate in weekly 90-min yoga class (yoga group) or a weekly 90-min free-choice physical exercise (control group). The study period lasted for 10 weeks. Prior to and at the end of the study period the participants completed a battery of self-reported inventories, including the Brief Symptom Inventory (BSI).

Results

Physical activity (including yoga) significantly reduced the inmates' levels of psychological distress. Yoga practice improved all primary symptom dimensions and its positive effect on the obsessive-compulsive, paranoid ideation, and somatization symptom dimensions of the BSI stayed significant even when comparing with the control group.

Conclusions

Yoga as a form of physical activity is effective for reducing psychological distress levels in prison inmates, with specific effect on symptoms such as suspicious and fearful thoughts about losing autonomy, memory problems, difficulty in making decisions, trouble concentrating, obsessive thought, and perception of bodily dysfunction.

INTRODUCTION

Originally, prisons were established to isolate convicted offenders in order to keep society safe. The major tasks of these institutions included punishment and correction of individuals who often had a history of aggression and/or antisocial behavior (1). Over time, the purpose of correctional institutions changed; it shifted from isolation and punishment to prevention of relapse into criminality by using different treatment programs. As psychiatric ill-health is a common condition among prison inmates (2, 3), today's correctional institutions pay greater attention to the issue of inmates' mental health as a factor determining the risk of relapse. It has been estimated that prisons hold more than three times as many persons with psychiatric problems and/ or diagnoses than can be found in psychiatric facilities (4). Personality and mood disorders, addiction, and neurodevelopmental syndromes are the most frequent mental health problems occurring in prison inmates (5, 6). Psychopathic personality profile (7), substance abuse and attention deficit hyperactivity disorder are the most recurring diagnoses when considering the risk of relapse into a criminal life style (8).

The Swedish Prison and Probation Service aims to enhance inmates' opportunities to reintegrate into society upon release and their ability to live a drug free and law-abiding life. To this end, the Prison and Probation Service requires all inmates to participate in rehabilitation activities, 6 h a day, 5 days a week. As the capabilities and needs of inmates differ greatly, there is a wide range of activities to choose from. Beside occupational activities, inmates are provided opportunities to participate in study programs, rehabilitation programs concerning drug- and alcohol abuse, violent behavior and aggression, yoga classes, health programs, drama, painting and writing classes, etc (9).

The Prison and Probation Service has offered yoga classes to inmates since 2008 [for details on the "Krimyoga" offered, please see Kerekes et al. (10)]. Yoga aims to reconstruct the balance between body and mind by using physical, psychological, and spiritual practices. The core of yoga is conscious, deep and slow breathing during which different body movements are performed, while shifting one's focus to the body in still positions (body awareness) and to one's feelings at the moment. In some yoga forms, the session ends with meditation. The physiological and psychological impacts of yoga have been tested in various samples: general population, clinical settings, and prison inmates (11–13). It has been shown that regular exercises of yoga reduce physiological response to stress (14), state and trait anxiety and depression (15), antisocial behaviors (10), as well as anger and aggression (16). Furthermore, regular yoga practice is coupled with increased positive and decreased negative emotional states (10, 17), decreased stress (17), as well as with significantly improved impulse control and sustained attention (10). Importantly, it has been shown that yoga in correctional settings has a positive effect on risk factors associated with criminal recidivism (10).

Beside the important psychological benefits, yoga practitioners experience physiological benefits, such as improved body posture and breathing technique, increased body awareness and easier relaxation (18). In terms of biological benefits, regular yoga exercise has been found to favor normalized activity of the hypothalamus, pituitary-, and adrenal glands, and the autonomic nervous system (19), to decrease the concentration of cortisol, and to increase the concentration of serotonin and melatonin (18). As yoga can be practiced in group settings, it can also have positive effects in terms of social relations and feelings of belonging among those practicing yoga together (20).

Although the effects of yoga have already been studied, only a few studies have been conducted among prison inmates. The aim of this study is to investigate the effect of 10 weeks of yoga practice on the participating prison inmates' mental health profile, operationalized in the form of psychological distress levels.

SUBJECTS AND METHODS

Procedure and Sample

The present study is part of a broader research, therefore for a detailed description of the study procedure and methods, please refer to Kerekes et al. (10). In short, they were as follows: During a 20-month period in 2013, 2014, and 2015, male and female inmates from several high- and medium-security class prisons participated in the study. After giving their consent to participate, the inmates were randomly placed in either of two groups: a group participating once a week for 10 weeks in a 90-min long yoga class (the yoga group); a group participating once a week for 10 weeks in a 90-min long free-choice non-yoga physical exercise (the control group). Prior to the start (Time 1) of the 10-week study period and again upon its completion (Time 2), the participants completed a survey, comprising several self-reported measures (assessing stress, aggression, affective states, sleep-quality, and psychological well-being). One of these measures was the Brief Symptom Inventory, which assesses a person's level of psychological distress.

A total sample of 226 inmates (201 men; 25 women) participated in the study. The average attrition rate was 32.7%. For detailed description of attrition rates and reasons in the yoga and control groups, please see Kerekes et al. (10). The final number of participants completing the whole study period was 152 (133 men; 19 women). The participants were evenly distributed between the yoga group and the control group [yoga group: 77 (67 male and 10 female) with a mean age of 36.4 years; control group: 75 (66 male and 9 female) with a mean age of 34.9 years]. A detailed description of the participants' characteristics, including comparison between the yoga and the control groups was previously published in Kerekes et al. (10), where participants in the yoga and control groups were comparable in gender distribution, age, length of sentence, and security level of holding prison.

Instrument—The Brief Symptom Inventory

The Brief Symptom Inventory (BSI) is a self-reported measure of psychological distress and psychiatric symptoms (21). The inventory consists of 53 items, assessing 9 primary symptom dimensions (see details below), and the Global Severity Index (GSI), which measures the individual's overall psychological distress level. Participants are asked to rate how much a problem has bothered or distressed them over the previous month. Items can be rated on a five-point Likert scale from 0 to 4 (0 = not at all, 1 = a little, 2 = moderately, 3 = substantially, and 4 = extremely). Higher scores therefore indicate greater psychological distress, whereas lower scores indicate better psychological well-being and/or psychiatric health. The 9 primary symptom dimensions are as follows:

1. Anxiety: Indications of tension, restlessness, and nervousness, as well as experiences of free-floating anxiety or panic. The following items defined the Anxiety scale in the present study:

   "Nervousness or shakiness inside," "suddenly scared for no reason," "feeling fearful," "feeling tense or keyed up," "spells of terror or panic," "feeling so restless you could not sit still."

2. Depression: A wide range of symptoms and signs from dysphoric affect and mood, loss of energy and interest in life activities to feelings of hopelessness and uselessness. The following items defined the Depression scale in the present study:

   "Thoughts of ending your life," "feeling lonely," "feeling blue," "feeling no interest in things," "feeling hopeless about the future," "feelings of worthlessness."

3. Interpersonal Sensitivity: Feelings of inadequacy and inferiority and signs of uneasiness and discomfort during interpersonal interactions. The following items defined the Interpersonal Sensitivity scale in the present study:

   "Your feelings are easily hurt," "feelings that people are unfriendly or dislike you," "feeling inferior to others," "feeling very self-conscious with others."

4. Hostility: Feelings of irritability and annoyance, the presence of frequent arguments and urges to break things. The following items defined the Hostility scale in the present study:

   "Feeling easily annoyed or irritated," "temper outbursts that you cannot control," "having urges to beat, injure, or harm someone," "having urges to break or smash things," "getting into frequent arguments."

5. Obsessive-Compulsive: Thoughts and actions that are experienced as irresistible and persistent. The following items defined the Obsessive-Compulsive scale in the present study:

   "Trouble remembering things," "feeling blocked in getting things done," "having to check and double check what you do," "difficulty in making decisions," "your mind going blank," "trouble concentrating."

6. Psychoticism: Psychoticism is described as a continuum pending between a mildly alien life style and extremely psychotic status. Outside the psychiatric population, this dimension can measure social alienation. The following items defined the Psychoticism scale in the present study:

   "The idea that someone else can control your thoughts," "feeling lonely even when you are with people," "never feeling close to another person," "the idea that something is wrong with your mind."

7. Paranoid Ideation: Suspicious, hostile, and fearful thoughts of losing autonomy. The following items defined the Paranoid Ideation scale in the present study:

   "Feeling that others are to blame for most of your troubles," "feeling that most people cannot be trusted," "feeling that you are watched or talked about by others," "others not giving you proper credit for your achievements," "feeling that people would take advantage of you if you let them."

8. Phobic Anxiety: Phobias such as fear of crowds, open spaces, conveyances, etc. The following items defined the Phobic Anxiety scale in the present study:

   "Feeling afraid in open spaces," "feeling afraid to travel on buses, subways, or trains," "having to avoid certain things, places or activities because they frighten you," "feeling uneasy in crowds," "feeling nervous when you are left alone."

9. Somatization: The perception of bodily dysfunction. Aches, pains, and other discomforts stemming from the musculature or complaints coupled to the cardiovascular, gastrointestinal, and respiratory systems. The following items defined the Somatization scale in the present study:

   "Faintness or dizziness," "pains in heart or chest," "nausea or upset stomach," "trouble getting your breath," "numbness or tingling in parts of your body," "feeling weak in parts of your body," "feeling heavy in arms or legs."

Yoga Classes and Free-Choice Physical Exercise

The yoga classes were led by prison officers who had received training in yoga instruction provided by the Swedish Prison and Probation Service. The yoga classes ("Krimyoga") were based on *hatha yoga*, a form of physical yoga that includes elements of relaxation specifically developed for use in correctional settings. Each yoga class comprised (in sequence) at least 10 min of warm up, a defined combination of *asanas* (yoga postures), about 5 min of breathing exercises, and finally 5 min of relaxation and deep relaxation. The weekly yoga classes lasted for 90 min and were carried out uniformly across the different correctional facilities. Free-choice physical exercise could typically include training in gym, walking, basketball, or football. Post-intervention assessment in our study included a question on time spent on physical activity and ensured that each participant (of the control group) indeed has spent at least 90 min every week with physical activity.

Data Analysis

As the low number of female participants in the study (19 of 152) did not allow gender-specific analyses, male and female subjects' data were analyzed together in this study. Violation of normality assumption was revealed by Shapiro–Wilk test in the assessed BSI data.

Non-parametric tests were therefore performed. Wilcoxon Signed Rank test was used to assess changes in BSI scores from Time 1 (beginning of the 10-week study period) to Time 2 (end of the study period) within each group (yoga and control) and Mann-Whitney *U*-test was conducted to compare differences between the groups. The effect size was calculated by dividing the z value by the square root of N (total number of cases). Cohen (22) criteria of 0.1 = small effect, 0.3 = medium effect, 0.5 = large effect were applied for the results interpretation. Significance level was defined at $p <$

0.05. The data was processed with the help of the statistical program SPSS version 23 (IBM).

Ethics

The regional Ethical Review Board in Linköping approved the study (Dnr 2013/302 – 31). The prison inmates received complete information (written and oral) about the study and the conditions of participation. Participation was entirely voluntary and all participants were ensured that their answers would in no way affect their actual prison sentence to be served. Participation required the signature of a written consent form. The collected information was treated confidentially and the researchers received only coded data. Participants who completed the whole study received a compensation of 200 Swedish crowns (about 20 euros/US dollars).

RESULTS

The data analysis revealed significant changes in all but three primary symptom dimensions (obsessive-compulsive, somatization, and phobic anxiety) in the control group. These significant changes were improvements with small effect sizes (r varied between 0.15 and 0.29) (Table 1). In the yoga group all dimensions of psychological distress were significantly improved with small or medium effect size (r varied between 0.19 and 0.36) (Table 1).

While the Global Severity index decreased in both groups with a similar, medium effect size, the effect size of the measured improvements in each primary dimension was always bigger within the yoga group. Table 1 summarizes the results of the within-group analyses on the General Severity Index (GSI) and the 9 primary symptom dimensions of the BSI in the control and the yoga groups separately.

Table 1. Changes within groups (yoga and control) measured with z-scores, means (M), and standard deviations (SD) on the global severity index and on the primary symptom dimensions at pre- and post-intervention assessment in both groups.

	Within groups											
	Control group (n = 75)						**Yoga group (n = 77)**					
Psychological distress BSI	**Time 1 M(SD)**	**Time 2 M(SD)**	**Average change**	p_{within}	z	r	**Time 1 M(SD)**	**Time 2 M(SD)**	**Average change**	p_{within}	z	r
Global severity index	1 (0.7)	0.8 (0.7)	−0.2 (0.4)	< 0.001	−3.7	0.30	0.8 (0.5)	0.6 (0.6)	−0.3 (0.5)	< 0.001	−3.8	0.30
Anxiety	1.2 (0.9)	0.9 (0.9)	−0.3 (0.7)	< 0.001	−3.62	0.29	1.1 (0.8)	0.8 (0.8)	−0.4 (0.6)	< 0.001	−4.69	0.33
Depression	1.2 (0.9)	1.0 (1.0)	−0.2 (0.8)	0.016	−2.41	0.19	1 (0.8)	0.7 (0.7)	−0.4 (0.7)	< 0.001	−4.68	0.33
Interpersonal Sensitivity	0.9 (0.9)	0.8 (1.0)	−0.1 (0.7)	0.046	−2	0.15	0.8 (0.8)	0.4 (0.6)	−0.3 (0.7)	< 0.001	−4.05	0.29
Hostility	0.9 (0.8)	0.7 (0.7)	−0.2 (0.8)	0.019	−2.34	0.18	0.8 (0.7)	0.5 (0.7)	−0.2 (0.6)	0.003	−3	0.21
Obsessive-compulsive	1.3 (0.9)	1.2 (0.9)	−0.2 (0.7)	0.057	−1.9	0.15	1.3 (0.9)	0.8 (0.7)	−0.4 (0.6)	< 0.001	−5.03	0.36
Psychoticism	0.9 (0.8)	0.7 (0.8)	−0.2 (0.6)	0.012	−2.5	0.20	0.7 (0.8)	0.5 (0.7)	−0.2 (0.7)	0.003	−2.95	0.21
Paranoid ideation	1.3 (0.9)	1 (0.9)	−0.2 (0.6)	0.005	−2.83	0.22	1.1 (0.9)	0.7 (0.7)	−0.5 (0.7)	< 0.001	−4.6	0.33
Phobic anxiety	0.6 (0.9)	0.5 (0.9)	−0.1 (0.7)	0.217	−1.23	0.09	0.6 (0.8)	0.3 (0.7)	−0.2 (0.6)	0.007	−2.7	0.19
Somatization	0.7 (0.6)	0.7 (0.6)	−0.1 (0.6)	0.114	−1.58	0.12	0.6 (0.6)	0.4 (0.6)	−0.3 (0.5)	< 0.001	−4.16	0.30

Between-group analyses revealed highly significant changes in three primary BSI symptom dimensions: obsessive-compulsive, paranoid ideation, and somatization. A significantly lower mean was reported for the score on the obsessive-compulsive dimension by the yoga group (mean rank = 62.72, n = 71) compared to the control group (mean rank = 82.01, n = 73), ($z = -2.79$, p = 0.005, r = 0.23). A significantly lower mean rank difference was also reported for the score on the paranoid ideation dimension by the yoga group (mean rank = 65.02, n = 71) compared to the control group (mean rank = 80.66, n = 74), ($z = -2.25$, p = 0.024, r = 0.18), and similarly for the primary dimension of somatization by the yoga group (mean rank = 64.46, n = 71) compared to the control group (mean rank = 80.32, n = 73), ($z = -2.29$, p = 0.022, r = 0.19). All of these differences had a small effect size. Table 2 summarizes the results of the between-group analyses on the General Severity Index (GSI) and the 9 primary symptom dimensions of the BSI.

Table 2. Between–group comparison on the global severity index and the primary symptom dimensions of BSI (*n* varies between 76 and 64 for the control group and between 77 and 50 for the yoga group in the different dimensions).

Psychological distress BSI	**Mean Rank**		p_{between}	z	r
	Control group	**Yoga group**			
Global severity index	61.28	52.66	0.17	−1.38	0.12
Anxiety	76.69	68.19	0.219	−1.23	0.10
Depression	78.91	67.63	0.105	−1.62	0.13
Interpersonal sen-sitivity	81.55	68.18	0.055	−1.92	0.16
Hostility	77.68	74.38	0.64	−0.47	0.04
Obsessive-compul-sive	82.01	62.72	0.005	−2.79	0.23
Psychoticism	75.81	72.11	0.593	−0.53	0.04
Paranoid ideation	80.66	65.02	0.024	−2.25	0.18
Phobic anxiety	77.94	69.18	0.191	−1.31	0.11
Somatization	80.32	64.46	0.022	−2.29	0.19

DISCUSSION

The present study provides further evidence that physical activity generally reduces psychological distress and psychiatric complaints in prison inmate samples. The main finding of the present study is that yoga practice has specific positive effects on the mental ill-health of prison inmates, offering significant help with symptoms of paranoid ideation, memory problems, trouble concentrating, obsessive thought, and somatization.

In this study, inmates were randomized into either a group where they could participate in a weekly yoga exercise or into a group where they could practice free-choice non-yoga physical activity, both during a 10-week period. Both groups of inmates reported significantly improved psychological distress levels, mostly in terms of decreased levels of anxiety, depressive symptoms, interpersonal sensitivity, suspiciousness, hostility, fearful thoughts, and social alienation.

General Effects of Physical Activity (Including Yoga) on the Level of Psychological Distress And/Or Psychiatric Complaints

This study revealed a significantly decreased level of anxiety and depressive symptoms in both groups. The negative association between physical activity and prevalence of depression and anxiety disorders has been repeatedly described in previous research and repeated now with our study. A few studies even examined the association between physical activity and these complaints in a prospective design. A review of these prospective studies suggests that physical activity may be clinically effective, at least in major depression and panic disorder (23). Moreover, cardiovascular and resistance training and high-intensity strength training have been found to significantly reduce depression scale scores after a 9-month testing period (24), cardiovascular and resistance training has showed significantly improved numbers in interpersonal sensitivity, and high-intensity strength training has showed improved numbers in anxiety, phobic anxiety, and hostility (24). This study supports that these measures (interpersonal sensitivity, phobic anxiety, and hostility) also significantly improve after 10 weeks of yoga exercises. Consistent with previous findings, this study also shows that yoga can be used advantageously against depression and anxiety (25, 26).

Previous research found that yoga, as a complement to drug treatment, is suitable for patients diagnosed with schizophrenia and helps patients with psychoticism or paranoid ideation to regain better quality of life and socio-professional function (26, 27). In this study, we found that these dimensions of psychological distress were significantly improved not only by yoga, but also by regular non-yoga free-choice physical activity.

Specific Effects of Yoga on Psychological Distress And/Or Psychiatric Complaints

In this study, we measured a significant decrease in paranoid symptoms—such as suspicious and fearful thoughts about others planning or willing to take one's autonomy—after 10 weeks of yoga classes when compared to the non-yoga control group. This is in line with previous findings from clinical studies where yoga therapy programs revealed significant improvement of both positive and negative psychotic symptoms in schizophrenia patients (28). This study is, however, the first to show a reduction of paranoid thoughts in prison inmates.

The other primary symptom dimension for which a significant difference was measured between the yoga group and the control group after the study period was the obsessive-compulsive symptom dimension. The effect of yoga particularly on obsessive-compulsive disorder was suggested in previous research (29). Furthermore, in a clinical population an increased effect of pharmaceutical treatment was observed when patients with obsessive-compulsive symptoms participated in yoga classes (30).

Yoga has a direct calming effect and can lead to an improved ability to control thoughts. Yoga may prevent or decrease thoughts from becoming obsessions and thus help the individual master his or her compulsive behavior (30). Positive results have been found for yoga practice as a complementary treatment for medication or psychotherapy for General Anxiety Disorder and Obsessive-Compulsive Disorder, including for imprisoned individuals (24). Moreover, significant improvements in concentration (31), sustained attention (10), and memory (32) have been measured after yoga practice.

This study also shown that yoga, but not non-yoga free-choice exercise, can significantly reduce symptoms of somatization. Yoga, in addition to its positive effects on mental conditions such as anxiety and depression, has also been found to have a positive effect on somatization symptoms (33), which is in accordance with our findings. Clinical symptoms of somatization include dizziness, headache, chest pain, nausea, difficulty breathing, feelings of body weakness, body numbness, and a heavy feeling in the body. Through the different variations of body postures the participants of yoga classes increase their body awareness, thus improving their sense of control over their own body (31), their perception of bodily dysfunction, and their ability to prevent psychosomatic symptoms (33).

The above discussed results of our study could be explained by underlining neurobiological changes. For example several studies have investigated the effects of yoga on the sympathetic and parasympathetic functions of the autonomic nervous system as well as on the regulation of the hypothalamus-pituitary-adrenal axis (29, 34). These studies show that regular yoga exercise can reduce the levels of cortisol and catecholamines (adrenaline and noradrenaline) and increase the levels of serotonin, melatonin and gamma-amino butyric acid, which are all important factors in the regulation of mental health. Other physiological effects of yoga practice, in particular improved cortisol levels, can also be associated with improved self-esteem and mental and emotional states. These yoga-induced chemical changes in the body entail improved experiences of well-being and reduced levels of psychological distress (34).

Strengths and Limitations

The present study was a randomized controlled trial. This method is used to minimize the bias effects of extraneous or irrelevant variables on the measurements, and provides the strongest evidence of the effects of a treatment. One limitation of the study is that the information about inmates' psychiatric health profile was operationalized and not assessed during a clinical examination. In addition, the use of self-reported measures is a limitation. Self-reported measures rely greatly on the respondent's capability to remember and admit the "truth"; answers may be distorted by social desirability and recall biases (35). However, self-reported research instruments present the significant advantages that they can be used without any clinical experience, are cost efficient, and can cover considerably larger numbers of individuals than when clinical measurements or semi-structured interviews are used for data collection in a given time frame. Furthermore, it has been found that in some cases it can be easier for respondents to report problems and complaints through an anonymous self-report, than during face-to-face clinical interview surveys or measures (35).

CONCLUSION AND SUGGESTIONS FOR FURTHER RESEARCH

This study confirms that physical activity has a positive effect on the psychological distress levels of prison inmates and provides new evidence about the specific positive effect of yoga on symptoms such as paranoid thoughts, memory problems, difficult decision making, trouble concentrating, obsessive thoughts, perception of bodily dysfunction, and somatization. The provision of opportunities for prison inmates to carry out physical activity in general and yoga exercise in particular can be an important tool for improving their reintegration into society and for helping them lead a drug free and law-abiding lifestyle upon release, which hypothesis should be tested in larger scale studies including follow-up recording of eventual long-term effects of the treatment.

AUTHOR CONTRIBUTIONS

AS was responsible for data analysis and interpretation, continuous progress monitoring, manuscript drafting and submission. PM and ST contributed to data analysis, interpretation, and manuscript drafting. NK designed and led the project and supervised the drafting and development of the manuscript.

Funding

This research was financed by the Swedish Prison and Probation Service with a project number 2012:251 and under the title: Psykobiologiska effekter av yoga i anstaltsmiljö.

Conflict of Interest Statement

NK was employed by the Swedish Prison and Probation Service during the study.

The remaining authors declare that the research was conducted in the absence of any commercial or financial relationships that could be construed as a potential conflict of interest.

ACKNOWLEDGMENTS

We thank the prison inmates for participating in this study and hope they have benefited from the experience. We wish to express our deep gratitude to the prison officers who acted as yoga teachers and provided valuable research assistance without which the study would not have been possible. We also thank Dr. Britt Hedman Ahlström for her insightful comments on the manuscript.

REFERENCES

1. Homel R, Thompson C. Causes and prevention of violence in prisons. In: O'Toole S, Eyland S, editors. *Corrections Criminology.* Sydney, NSW: Hawkins Press (2005). p. 101–8.
2. Slade K, Samele C, Valmaggia L, Forrester A. Pathways through the criminal justice system for prisoners with acute and serious mental illness. *J Forensic Leg Med.* (2016) 44:162–8. doi: 10.1016/j.jflm.2016.10.007
3. Hofvander B, Anckarsäter H, Wallinius M, Billstedt E. Mental health among young adults in prison: the importance of childhood-onset conduct disorder. *Br J Psychiatr Open* (2017) 3:78–84. doi: 10.1192/bjpo.bp.116.003889
4. Markowitz FE. Mental illness, crime, and violence: risk, context, and social control. *Aggress Violent Behav.* (2011) 16:36–44. doi: 10.1016/j.avb.2010.10.003
5. Fazel S, Danesh J. Serious mental disorder in 23000 prisoners: a systematic review of 62 surveys. *Lancet* (2002) 359:545–50. doi: 10.1016/S0140-6736(02)07740-1
6. Sansone RA, Sansone LA. Borderline personality and criminality. *Psychiatry* (2009) 6:16–20.
7. PubMed Abstract | Google Scholar
8. Moffitt TE, Caspi A, Harrington H, Milne BJ. Males on the life-course-persistent and adolescence-limited antisocial pathways: Follow-up at age 26 years. *Dev Psychopathol.* (2002) 14:179–207. doi: 10.1017/S0954579402001104
9. Mohr-Jensen C, Steinhausen HC. A meta-analysis and systematic review of the risks associated with childhood attention-deficit hyperactivity disorder on long-term outcome of arrests, convictions, and incarcerations. *Clin Psychol Rev.* (2016) 48:32–42. doi: 10.1016/j.cpr.2016.05.002
10. Kriminalvården. *Rehabilitation - Swedish Prison and Probation Service | Kriminalvården* (2018) Available online at: https://www.kriminalvarden.se/swedish-prison-and-probation-service/rehabilitation/ (Accessed April 21, 2018).
11. Kerekes N, Fielding C, Apelqvist S. Yoga in correctional settings: a randomized controlled study. *Front Psychiatry* (2017) 8:204. doi: 10.3389/fpsyt.2017.00204

12. Sharma M, Haider T. Yoga as an alternative and complementary therapy for patients suffering from anxiety. *J Evid Based Complement Altern Med.* (2013) 18:15–22. doi: 10.1177/2156587212460046
13. Kahya HH, Raspin CG. Yoga therapy for the mind eight-week course: participants' experiences. *Explor J Sci Heal.* (2017) 13:116–23. doi: 10.1016/j.explore.2016.12.006
14. Streeter CC, Gerbarg PL, Saper RB, Ciraulo DA, Brown RP. Effects of yoga on the autonomic nervous system, gamma-aminobutyric-acid, and allostasis in epilepsy, depression, and post-traumatic stress disorder. *Med Hypotheses* (2012) 78:571–9. doi: 10.1016/j.mehy.2012.01.021
15. Rani K, Tiwari S, Singh U, Agrawal G, Ghildiyal A, Srivastava N. Impact of yoga nidra on psychological general wellbeing in patients with menstrual irregularities: a randomized controlled trial. *Int J Yoga* (2011) 4:20–5. doi: 10.4103/0973-6131.78176
16. Michalsen A, Jeitler M, Brunnhuber S, Lüdtke R, Büssing A, Musial F, et al. Iyengar yoga for distressed women: a 3-armed randomized controlled trial. *Evid Based Complement Altern Med.* (2012) 2012:408727. doi: 10.1155/2012/408727
17. Raghuram N, Deshpande S, Nagendra H. A randomized control trial of the effect of yoga on verbal aggressiveness in normal healthy volunteers. *Int J Yoga* (2008) 1:76–82. doi: 10.4103/0973-6131.41034
18. Bilderbeck AC, Farias M, Brazil IA, Jakobowitz S, Wikholm C. Participation in a 10-week course of yoga improves behavioural control and decreases psychological distress in a prison population. *J Psychiatr Res.* (2013) 47:1438–45. doi: 10.1016/j.jpsychires.2013.06.014
19. Klatte R, Pabst S, Beelmann A, Rosendahl JS. The efficacy of body-oriented yoga in mental disorders. *Dtsch Arztebl Int.* (2016) 113:195–202. doi: 10.3238/arztebl.2016.0195
20. Jeter PE, Slutsky J, Singh N, Khalsa SBS. Yoga as a therapeutic intervention: a bibliometric analysis of published research studies from 1967 to 2013. *J Altern Complement Med.* (2015) 21:586–92. doi: 10.1089/acm.2015.0057
21. de Manincor M, Bensoussan A, Smith CA, Barr K, Schweickle M, Donoghoe LL, et al. Individualized yoga for reducing depression and anxiety, and improving well-being: a randomized controlled trial. *Depress Anxiety* (2016) 33:816–28. doi: 10.1002/da.22502

22. Derogatis LR. The brief symptom inventory: an introductory report. *Psychol Med.* (1983) 13:595–605. doi: 10.1017/S0033291700048017

23. Cohen J. *Statistical Power Analysis for the Behavioral Sciences.* 2nd ed. Hillsdale, NJ: Erlbaum (1988).

24. Ströhle A. Physical activity, exercise, depression and anxiety disorders. *J Neural Transm.* (2009) 116:777–84. doi: 10.1007/s00702-008-0092-x

25. Battaglia C, Di Cagno A, Fiorilli G, Giombini A, Borrione P, Baralla F, et al. Participation in a 9-month selected physical exercise programme enhances psychological well-being in a prison population. *Crim Behav Ment Heal.* (2015) 25:343–54. doi: 10.1002/cbm.1922

26. Tolbaños Roche L, Miró Barrachina MT, Ibáñez Fernández I. Effect of "Exercise Without Movement" yoga method on mindfulness, anxiety and depression. *Complement Ther Clin Pract.* (2016) 25:136–41. doi: 10.1016/j.ctcp.2016.09.008

27. Balasubramaniam M, Telles S, Doraiswamy PM. Yoga on our minds: a systematic review of yoga for neuropsychiatric disorders. *Front Psychiatry* (2013) 3:117. doi: 10.3389/fpsyt.2012.00117

28. Vancampfort D, De Hert M, Vansteenkiste M, De Herdt A, Scheewe TW, Soundy A, et al. The importance of self-determined motivation towards physical activity in patients with schizophrenia. *Psychiatry Res.* (2013) 210:812–8. doi: 10.1016/j.psychres.2013.10.004

29. Visceglia E, Lewis S. Yoga therapy as an adjunctive treatment for schizophrenia: a randomized, controlled pilot study. *J Altern Complement Med.* (2011) 17:601–7. doi: 10.1089/acm.2010.0075

30. Kirkwood G, Rampes H, Tuffrey V, Richardson J, Pilkington K, Ramaratnam S. Yoga for anxiety: a systematic review of the research evidence. *Br J Sports Med.* (2005) 39:884–91; discussion: 891. doi: 10.1136/bjsm.2005.018069

31. Bhat S, Varambally S, Karmani S, Govindaraj R, Gangadhar BN. Designing and validation of a yoga-based intervention for obsessive compulsive disorder. *Int Rev Psychiatry* (2016) 28:327–33. doi: 10.3109/09540261.2016.1170001

32. Javnbakht M, Hejazi Kenari R, Ghasemi M. Effects of yoga on depression and anxiety of women. *Complement Ther Clin Pract.* (2009) 15:102–4. doi: 10.1016/j.ctcp.2009.01.003

33. Rocha KKF, Ribeiro AM, Rocha KCF, Sousa MBC, Albuquerque FS, Ribeiro S, et al. Improvement in physiological and psychological parameters after 6months of yoga practice. *Conscious Cogn.* (2012) 21:843–50. doi: 10.1016/j.concog.2012.01.014
34. Yoshihara K, Hiramoto T, Oka T, Kubo C, Sudo N. Effect of 12 weeks of yoga training on the somatization, psychological symptoms, and stress-related biomarkers of healthy women. *Biopsychosoc Med.* (2014) 8:1. doi: 10.1186/1751-0759-8-1
35. Woolery A, Myers H, Sternlieb B, Zeltzer L. A yoga intervention for young adults with elevated symptoms of depression. *Altern Ther Health Med.* (2004) 10:60–3.
36. Nijman H, Bjørkly S, Palmstierna T, Almvik R. Assessing aggression of psychiatric patients: methods of measurement and its prevalence. In: Richter D, Whittington R, editors. *Violence in Mental Health Settings - Causes, Consequences, Management*. New York, NY: Springer (2006). p. 11–23.

INDEX

H

I

L

M

N

O